To my mother and father

FORMAL KNOT THEORY

by Louis H. Kauffman

Mathematical Notes 30
Princeton University Press
1983

Copyright © 1983 by Princeton University Press

All Rights Reserved

Printed in the United States of America by
Princeton University Press, 41 William Street,
Princeton, New Jersey 08540

ISBN 0-691-08336-3

The Princeton Mathematical Notes are edited by William Browder,
Robert Langlands, John Milnor, and Elias M. Stein

Libary of Congress Cataloging in Publication Data
will be found on the last printed page of this book

CONTENTS

		page
1.	Introduction...................................	1
2.	States, Trails, and the Clock Theorem..............	12
3.	State Polynomials and the Duality Conjecture.......	53
4.	Knots and Links...................................	67
5.	Axiomatic Link Calculations.......................	78
6.	Curliness and the Alexander Polynomial............	95
7.	The Coat of Many Colors...........................	105
8.	Spanning Surfaces................................	114
9.	The Genus of Alternative Links....................	125
10.	Ribbon Knots and the Arf Invariant................	143
Appendix. The Classical Alexander Polynomial...........		156
References..		165

1. Introduction

These notes constitute an exploration in combinatorics and knot theory. Knots and links in three-dimensional space may be understood through their planar projections. A knot is usually drawn as a schematic snapshot, with crossings indicated by broken line segments. Thus ⟨trefoil⟩ represents the trefoil knot. We shall refer to such a picture as a <u>knot diagram</u>. The projection corresponding to such a diagram forms a (directed) multi-graph in the plane, with four edges incident to each vertex. We shall study these graphs separately, and for their own sake.

In order to do this work it is very important to underline key concepts by adopting terminology and conventions that are easy to remember. For this reason I have taken a perhaps startling, but certainly memorable, set of terms for the graph - theoretic side of the ledger.

A (directed) planar graph with four edges incident to each vertex will be termed a <u>universe</u>. Thus ⟨trefoil⟩ is a trefoil universe. These universes have <u>singularities</u> (the crossings); they will also have <u>states</u>, <u>black holes</u>, <u>white holes</u>, and <u>stars</u>.

A <u>state</u> of a universe is an assignment of one <u>marker</u> per vertex in the forms

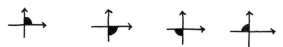

so that each region in the graph receives no more than one marker. Thus ⬬ is a state of the trefoil universe. Two regions of the state will be free of markers (since the number of regions exceeds the number of vertices by two in a connected universe). These free regions are inhabited by the <u>stars</u>. (∗).

It should be mentioned at once that each universe has states. In fact, <u>states with stars in adjacent regions are in one-to-one correspondence with Jordan trails on the universe</u>. A <u>Jordan trail</u> is an (unoriented) path that traverses every edge of the universe once and forms a simple closed curve in the plane. This correspondence is obtained by regarding each state marker as an instruction to <u>split</u> its crossing according to the following schema:

By splitting all crossings in a state, the Jordan trail automatically appears. Conversely, a choice of stars at the Jordan trail determines a specific state. (See section 2. for more details.) The process is illustrated below for the trefoil.

This correspondence underlines the importance of states with adjacent stars; further reference to states will assume star adjacency unless otherwise specified. The state markers are classified into the categories <u>black holes</u>, <u>white holes</u>, <u>up</u>, <u>down</u> according to placement with respect to the crossing orientation:

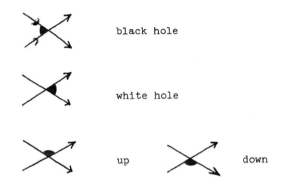

The <u>sign of a state</u> S, $\sigma(S)$, is defined by the formula $\sigma(S) = (-1)^b$ where b denotes the number of black holes in S. Just as the sign of a permutation changes under single transpositions of its elements, so does the sign of a state change under a <u>state transposition</u>. A state transposition is a move from one state to another that is obtained by switching a pair of state markers as indicated below.

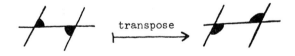

Note that in a state transposition both state markers **rotate by one quarter turn in the same clock-direction**. A state transposition in which the markers turn clockwise (counter-clockwise) will be termed a <u>clockwise</u> (<u>counter-clockwise</u>) move. A state is said to be <u>clocked</u> if it admits only clockwise moves, <u>mixed</u> if it admits both clockwise and counter-clockwise moves, and <u>counter-clocked</u> if it admits only counter-clockwise moves.

The key combinatorial result in our study is the following theorem.

<u>The Clock Theorem</u> (2.5). Let U be a universe and $\mathcal{S}$ the set of states of U for a given choice of adjacent fixed stars. Then $\mathcal{S}$ has a unique clocked state and a unique counter-clocked state. Any state in $\mathcal{S}$ can be reached from the clocked (counter-clocked) state by a series of clockwise (counter-clockwise) moves. Hence any two states in $\mathcal{S}$ are connected by a series of state transpositions.

By defining $S < S'$ whenever there is a series of clockwise moves connecting the state S' to the state S, the collection of states becomes a lattice whose top is the clocked state, and whose bottom is the counter-clocked state.

This result is illustrated for a particular universe in Figure 1. An oriented universe has directed edges so that each crossing has the form ⤫ .

In section 2. we shall give an algorithm for constructing the clocked and counter-clocked states. This algorithm, to-

5

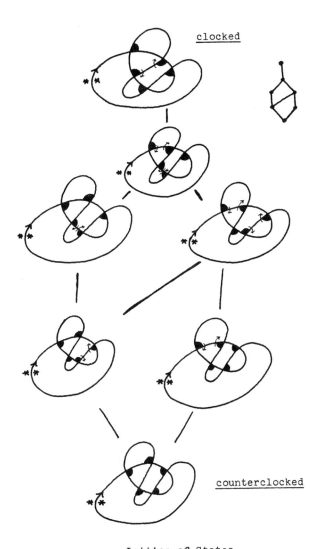

Lattice of States

Figure 1

gether with the Clock Theorem, gives an efficient method for enumerating all the states of a universe.

Patterns of black and white holes in the states give rise to a <u>Duality Conjecture</u> and to a series of results bridging combinatorics and the topology of knots and links.

<u>Duality Conjecture</u>. Let $\mathcal{S}$ be the collection of states of an oriented universe U with a choice of fixed adjacent stars. Let $N(r,s,\mathcal{S}) = N(r,s)$ denote the number of states in $\mathcal{S}$ with r black holes and s white holes. I conjecture that $N(r,s) = N(s,r)$ for all r , s.

As we shall see in section 3. <u>$N(r,s)$ is independent of star placement</u>. That is, if $\mathcal{S}'$ is another state collection arising from a different choice of fixed stars, then $N(r,s,\mathcal{S}) = N(r,s,\mathcal{S}')$ for all r , s.

This independence result depends crucially and subtley on the Clock Theorem. It is verified by interpreting the <u>state polynomial</u> $F(\mathcal{S}) = \sum_{r,s} (-1)^r N(r,s,\mathcal{S}) B^r W^s$ (belonging to the polynomial ring in variables B and W over the integers: $Z[B,W]$) as a determinant of a matrix associated with the universe and with $\mathcal{S}$. The signs of the permutations that occur in the expansion of this determinant coincide with the signs $(-1)^r$, and these are the signs of the states being enumerated!

A generalization of the state polynomial marks the transition into the theory of knots. A knot-diagram is a universe with extra structure at the crossings. To create a knot-diagram from a given universe entails a two-fold choice at each crossing. Hence 2^c knot-diagrams project to a universe with c crossings. It is convenient to designate these choices by placing a _code_ at each crossing. Our codes take the following forms:

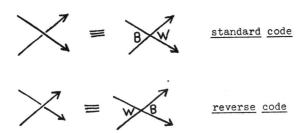

standard code

reverse code

Thus a knot or link diagram is an oriented universe with standard or reverse codes at each crossing.

In standard code the labels B and W hover over potential black and white holes respectively. Labels are flipped in the reverse code. The knot or link obtained by labelling a universe entirely with standard (reverse) code will be called a _standard_ (_reverse_) _knot_. A reverse knot is the mirror image of the corresponding standard knot. The trefoil is standard.

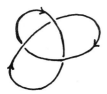 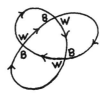

Let K be a knot and S a state, both sharing the same underlying universe U. We define an __inner-product__ $\langle K|S \rangle \in Z[B,W]$ and a __state polynomial__ $\langle K|\mathcal{S} \rangle$,

$$\langle K|\mathcal{S}\rangle = \Sigma_{S \in \mathcal{S}} \langle K|S \rangle$$

so that when K is standard this new polynomial coincides with the original state polynomial for $\mathcal{S}$. In order to do this the inner product is defined as follows:

__Superimpose__ K and S on the universe U. Let x denote the number of coincidences of W - labels in K with state markers in S. Let y denote the number of coincidences of B - labels with state markers. Then

$$\langle K|S \rangle = \sigma(S) W^x B^y.$$

When K is standard, x is the number of white holes and y is the number of black holes in S. For example, if K is the trefoil knot, then

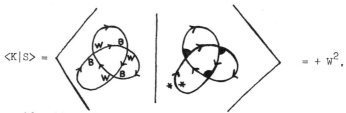

and consideration of two other trefoil states shows that $\langle K|\mathcal{S} \rangle = B^2 - WB + W^2$.

These state polynomials give rise at once to topological invariants of the knot or link K. To obtain a polynomial that is a topological invariant, __simply set__ $\underline{WB = 1}$ __and let__

$z = W - B$. Then $\langle K|\mathcal{S}\rangle$ becomes a polynomial in z, $\nabla_K(z)$, and $\nabla_K(z)$ is a topological invariant of K. In the trefoil example we have $\nabla_K = 1 + z^2$.

The polynomial ∇_K is identical with what I called the <u>Alexander-Conway polynomial of K</u>. It is a refinement of the classical Alexander polynomial ([1], [5], and [21]) and is characterized by the following three axioms:

1. To each oriented knot or link K there is associated a polynomial $\nabla_K(z) \in Z[z]$ such that ambient isotopic links receive identical polynomials.

2. If K is an unknotted circle then $\nabla_K = 1$.

3. If K, $\overline{K}$, and L are three links that differ at the site of <u>one</u> crossing as indicated below, then
$\nabla_K - \nabla_{\overline{K}} = z\nabla_L$.

K $\overline{K}$ L

These axioms alone suffice to calculate the polynomial, without reference to the underlying state polynomial or to any other model. See section 5. of these notes for a self-contained treatment via the axioms. The Alexander-Conway polynomial is a true refinement of the Alexander polynomial. Because it is defined absolutely (rather than up to sign and powers of the variable) it is capable of distinguishing many links from their mirror images - a capability not available to the Alexander polynomial.

The model using the state polynomial is a revision of Alexander's original combinatorial appraoch. The Clock Theorem provides the underpinning that makes this sign - precision possible - by allowing the identification of state signs with permutation signs (see section 3). In section 6. we detail the connection with the Alexander polynomial and show how it contains a hidden calculation of the Whitney degree ([14], [38]) of the knot projection seen as a plane curve immersion. In this section a combinatorial version of the Whitney degree, dubbed <u>curliness</u>, is given for universes.

Section 7. develops the multi-variable link polynomials via state polynomial models. Here the power of our method really comes into play since there is, at present, no axiomatization for these polynomials. The state summation method then gives simple proofs of John Conway's identities for the "polychrome skein".

Section 8. constitutes a necessary digression into the topology of orientable spanning surfaces for knots and links. We discuss genus, Seifert surface, Seifert pairing, and another model of the Alexander-Conway polynomial.

Section 9. contains our main topological theorem, a wide generalization of the Crowell-Murasugi Theorem on the genus of alternating knots ([8], [29]). I generalize this theorem to the class of <u>alternative</u> <u>knots</u> (<u>links</u>). Alternatives include alternating knots and links, knots that arise as the links of plane algebroid singularities ([27]), the Lorenz links of Birman and Williams ([3]), and many others. Our proof of the Theorem is quite simple - thanks going once again to the Clock Theorem for control of signs!

Section 10. discusses the Arf invariant of a knot. We show that the Arf invariant is the mod-2 reduction of the degree-2 coefficient of the Alexander-Conway polynomial. Then, with the help of $\sqrt{-1}$, we obtain a quick proof of J. Levine's Theorem relating the Arf invariant and the value of the Alexander polynomial at -1 taken modulo eight.

The appendix is an exposition of a classical approach to the Alexander polynomial via the Dehn presentation of the knot group.

I would like to take this opportunity to thank the many people with whom I have worked and conversed over these many years. Particular thanks go to my collaborators Tom Banchoff, Alan Durfee, Deborah Goldsmith, Walter Neumann, Larry Taylor, and Francisco Varela; to George Spencer-Brown for conversation, inspiration and insight into formal mathematics; to Kyoko Inoue and David Solzman for innumerable informal conversations; and to Milton Kerker, William Browder, and Ralph Fox for introducing me to research and to the study of knots. Finally, my thanks to Ms. Shirley Roper, Head Math Typist at the University of Illinois at Chicago, for her excellent typing job.

June 9, 1982
Chicago, Illinois

2. States, Trails, and The Clock Theorem.

In this section I shall delineate the states - trails correspondence, give an algorithm for constructing clocked (counterclocked) states, and prove the Clock Theorem. It is assumed that the reader is familiar with these concepts from the introduction, but, for the sake of completeness, exact definitions will be given below. Before embarking on this formal development, I would like to make a few remarks that highlight the key ingredients in the proof of the Clock Theorem: A universe in the form given in Figure 2 will be called a shell composition. It is quite clear that any shell composition has a unique clocked state where each shell receives markers in the form

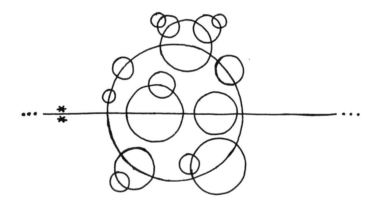

Figure 2

Note that the markers within this elementary shell stay within the shell when they rotate clockwise. If a shell composition is elaborated to a more complicated universe, then markers interior to the original shells may have the opportunity to rotate outwards. If the original shell is clocked, then these new rotations are necessarily counter clockwise. <u>Interior clocking leads to the potential for exterior counter clockwise moves</u>.

For example, suppose that 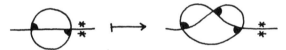 is elaborated to the form . The new universe has a state with a counter clockwise move derived from an interior shell marker rotating outward.

On the other hand, there are elaborations of shell compositions such as

where the new state is still clocked. We shall see that these properties of shell compositions are central to the Clock Theorem. The unique clocked and counter-clocked states of a universe are found by deriving an appropriately related shell composition.

States and Trails

<u>Definition</u> <u>2.1</u>. A <u>universe</u> is a connected planar (multi-) graph with 4-valent vertices. A universe is <u>oriented</u> if each

edge in the graph is directed so that the vertices take the form of an oriented crossing of two line segments:

We distinguish two methods of traversing a universe in the neighborhood of one of its vertices. A path that proceeds across a vertex along one of the aforementioned line segments will be said to <u>cross</u> that vertex. A path that meets a vertex but does not cross it will be said to <u>call</u> that vertex. Note that paths may move against the orientation directions of the universe. There are two possible local calling configurations at each vertex:

crossing   sites

Each of these consists in two segments of the path meeting at the vertex. It is convenient to separate these segments to form a <u>site</u>. As indicated above, a site is drawn so that each segment has a cusp point. If these cusps are brought together, the site becomes a crossing. An exchange of sites

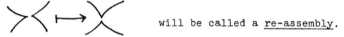 will be called a <u>re-assembly</u>.

A path that traverses the entire universe, using each edge once and calling every vertex, will be called a <u>Jordan trail</u>. A Jordan trail becomes a Jordan curve in the plane if we separate the path at each vertex to form a site. Such Jordan curves with sites will always be used to indicate

Jordan trails.

Lemma 2.2. Every universe admits at least one Jordan trail.

Proof. It follows from the Jordan curve theorem that a universe cannot be disconnected by both of the possible splittings of a given vertex. Therefore, choose a sequence of vertex splits that retain connectivity at every stage, and so that every vertex is split. The graph with sites that results from this process is a simple closed curve. Hence it corresponds to a Jordan trail on the universe.

The method of this proof is illustrated in Figure 3.

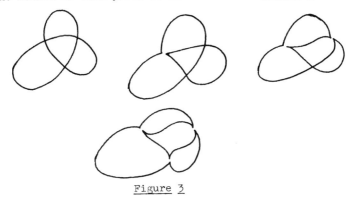

Figure 3

In closing the sites of a trail we obtain a universe, but forget how it was brought about. In order to remember how to split the vertices and regain the trail, place a state marker of the form 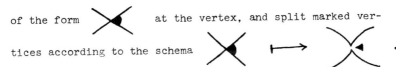 at the vertex, and split marked vertices according to the schema

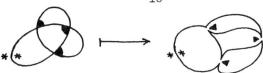

Definition 2.3. A <u>state</u> of a universe U is an assignment of state markers to vertices so that no region of U contains more than one state marker. Unless otherwise specified, it is assumed that the (two) unoccupied regions are adjacent and marked by <u>stars</u> (∗).

Theorem 2.4. Let $\mathcal{S}$ denote the collection of all states of a universe U that share a fixed choice of adjacent stars. Let $\mathcal{T}$ be the collection of all Jordan trails on U. Then $\mathcal{S}$ and $\mathcal{T}$ are in one-to-one correspondence.

<u>Proof</u>. Given a state S, let a(S) denote the result of splitting all vertices as specified by their state markers. In the step-by-step process of vertex splits we can never arrive at a situation where the marked and unmarked (m and n respectively in ⟩⟨) sides of a vertex are part of the same region. For if this were so, then this region would separate the stars (and the stars are adjacent). This guarantees the loss of one region after each vertex is split. Hence there must be exactly two regions that remain after all vertices are split. Thus a(S) is a Jordan trail, and we have $a : \mathcal{S} \longrightarrow \mathcal{T}$.

To obtain $b : \mathcal{T} \longrightarrow \mathcal{S}$, grow two trees, each rooted at one star as indicated in Figure 4. These trees then determine a collection of state markers so that ab and ba are identity maps. This completes the proof.

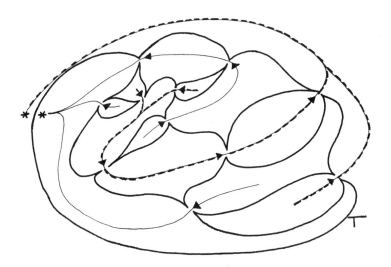

Tree growth creates state b(T) from trail T.

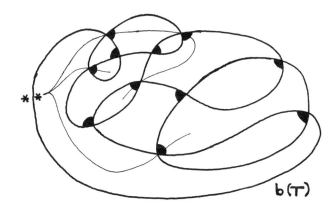

Figure 4

Remark. A checkerboard coloring of the universe divides the regions into those that lie inside and those that lie outside any Jordan trail.

Thinking of states as region - vertex assignments, hence as permutations of the vertices relative to an ordering of the regions, it is natural to consider transposition of permutations. The geometric version of a transposition is indicated in Figure 5. In this figure the regions X and Y are distinct, and the diagrams show part of a state with adjacent stars. The regions share common boundary that abuts to two state markers, one from each region. Under these conditions, it is possible to switch the markers as shown in Figure 5. A new state is formed by this switch. Call such a move a (state) transposition.

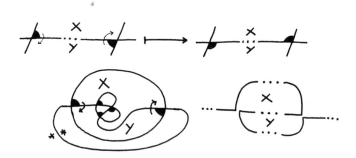

Figure 5

Figure 5 also shows sites that correspond to the transposition. Dotted lines display how the rest of the trail connects with this local situation. While there will in general be many other sites in such a trail, the topology of the connection to the transposition sites will always be as indicated.

For states with adjacent stars it is easy to verify that both markers in the transposition rotate in the same clock direction. Hence transpositions may be labelled <u>clockwise</u> or <u>counterclockwise</u> accordingly.

A state is said to be <u>clocked</u> if it admits only clockwise transpositions and <u>counterclocked</u> if it admits only counterclockwise transpositions. We aim to prove the

<u>Clock Theorem</u> 2.5. Let U be a universe and $\mathcal{S}$ the set of states of U for a given choice of adjacent fixed stars. Then $\mathcal{S}$ has a unique clocked state and a unique counterclocked state. Any state in $\mathcal{S}$ can be reached from the clocked (counterclocked) state by a series of clockwise (counterclockwise) moves. Hence any two states in $\mathcal{S}$ are connected by a series of state transpositions.

By defining $S < S'$ whenever there is a series of clockwise moves connecting the state S' to the state S, the collection of states becomes a lattice whose top is the clocked state, and whose bottom is the counterclocked state.

An important consequence of the Clock Theorem is that each state S has an intrinsically defined <u>sign</u>, $\sigma(S)$, that may be identified in a coherent way with the sign of an associated permutation. (In order to see this we first classify the state markers into the categories <u>black holes</u>, <u>white holes</u>, <u>up</u>, and <u>down</u> as in the introduction.)

<u>Definition</u> <u>2.6</u>. Let S be a state of an oriented universe U. Let $b = b(S)$ denote the number of black holes in S. Define the <u>sign of the state</u> S by the formula $\sigma(S) = (-1)^{b(S)}$.

<u>Definition</u> <u>2.7</u>. Let U be an oriented universe with regions $R_1, R_2, \ldots, R_n, R_{n+1}, R_{n+2}$ and vertices $V_1, V_2, \ldots, V_n$. Let $\mathcal{S}$ be the collection of states of U with stars in R_{n+1} and R_{n+2} (letting these denote adjacent regions). Choose the ordering of the vertices so that the region - vertex assignment $(R_i - V_i \;/\; i = 1, \ldots, n)$ corresponds to a state S_0 in $\mathcal{S}$, and so that this state has sign equal to 1. Let $S(n)$ denote the set of permutations of the letters $1, 2, \ldots, n$. For any permutation p in $S(n)$, let $sgn(p)$ denote the sign of p. Recall that $sgn(p)$ is equal to $(-1)^t$ where t is the number of moves in a sequence of transpositions that transforms p to the identity permutation $e = 123\ldots n$.

Since each state S in $\mathcal{S}$ is completely determined by some region - vertex assignment, there is a well-defined injection $P: \mathcal{S} \longrightarrow S(n)$ where $V_{P(S)(i)}$ has its marker in region R_i ($i = 1, \ldots n$). Call $P: \mathcal{S} \longrightarrow S(n)$ a <u>permutation assignment for</u> $\mathcal{S}$.

Lemma 2.7. Let S and S' be states of an oriented universe so that S' is obtained from S by <u>one</u> state transposition. Then S and S' differ in sign: $\sigma(S') = -\sigma(S)$.

Proof. Contemplate Figure 6. This completes the proof.

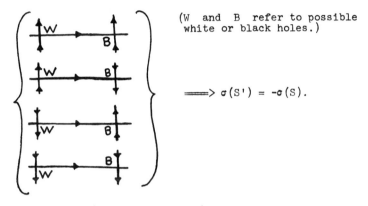

(W and B refer to possible white or black holes.)

$\Longrightarrow \sigma(S') = -\sigma(S).$

Figure 6

Proposition 2.8. Let $\mathcal{S}$ be a state collection for an oriented universe U, and $P: \mathcal{S} \longrightarrow S(n)$ a permutation assignment for $\mathcal{S}$. Then the signs of the states agree with the signs of their corresponding permutations. That is, $\sigma(S) = \text{sgn}(P(S))$ for all S in $\mathcal{S}$.

Proof. Use the notation of Definition 2.7. Thus S_0 is a state with $\sigma(S_0) = \text{sgn}(P(S_0)) = 1$. Let $e = P(S_0)$. Let S be any state in $\mathcal{S}$. Then, by the Clock Theorem, S can be obtained from S_0 by a sequence of state transpositions. Let t be the number of transpositions in some such sequence. Then, since P transforms state transpositions into transpositions of permutations, $\text{sgn}(P(S)) = (-1)^t$, while $\sigma(S) = (-1)^t$ by Lemma 2.7. Thus $\sigma(S) = \text{sgn}(P(S))$, completing the proof.

This agreement between state and permutation signs will be of great use to our further investigation of states in section 3., and to the ascent into knot theory that begins in section 4. In order to prove the Clock Theorem we first discuss an algorithm for constructing clocked and counter-clocked states.

Constructing Extremal States

By an <u>extremal</u> <u>state</u> I mean either a clocked state or a counterclocked state. It will be most convenient to phrase the discussion in terms of universes in <u>string form</u>; these will be called <u>strings</u>. A string is obtained by deleting an open interior arc from an edge of a universe. Thus

 is a trefoil string. It will be assumed that any state of a string corresponds to a state whose stars are in the two regions adjacent to the deleted edge. Thus

is a state of the trefoil string.

Drawing in this manner, the starred regions are always above and below the string.

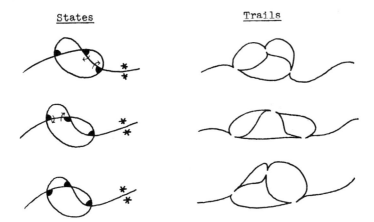

Here we indicate the three trefoil states (and available clockwise moves on each state). The corresponding trail is drawn to the right of each state.

Before giving formal details of the clocking algorithm, here is an example:

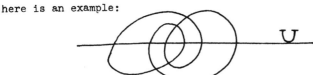

In order to make the outer trail boundary appear in the (clocked) form 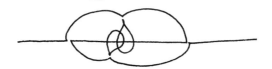, trace U as follows:

Tracing the outer shell leaves a similar configuration on the outside. Repeat the procedure on the inner form (recursively).

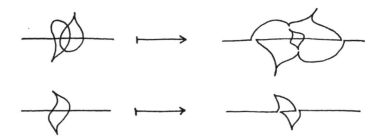

Putting this all together, obtain the trail T

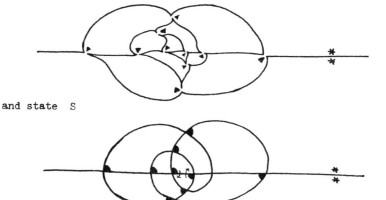

and state S

Here S is well wound up, with only one available clockwise move.

Definition 2.9. Let A and B be strings as below, and define A ⊕ B by splicing the right line of A to the left line of B (respecting orientations if these are present).

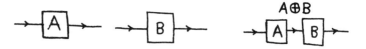

A string C is said to be irreducible if it cannot be written in the form $A \oplus B$ unless $A = E$ or $B = E$ where E denotes the trivial string ─→─. We shall refer to the left and right lines of the string as the <u>input line</u> and <u>output line</u> respectively.

<u>Remark</u>. Any string determines two universes with specific starred regions. These universes are obtained by connecting the input to the output above or below the body of the string. We may, if we like, regard the input line of the string as extending indefinitely to the left, and the output as extending indefinitely to the right. Then this (infinite) string divides the plane into two unbounded regions that correspond to the starred regions in either of the related universes. With this convention, we shall refer to the <u>bounded</u> and <u>unbounded regions of the string</u>.

<u>Definition 2.10</u>. An edge in a string A is an <u>interior edge</u> of A if e separates two bounded regions in A. An edge e is a <u>connecting edge</u> if it separates the two unbounded regions.

<u>Definition 2.11</u>. Let A and B be given strings, and p an interior point of a non-connecting edge of A. Let $A \oplus [B,p] \equiv A \oplus [B] = C$ denote the string obtained by replacing the trivial string at p by a copy of B (that does not intersect the rest of A).

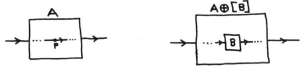

We shall say that $C = A \oplus [B]$ decomposes into a <u>carrier</u> A and a <u>rider</u> B. This definition generalizes to a collection of riders $\{B_1, B_2, \ldots, B_n\}$ and a set of points $\{p_1, p_2, \ldots, p_n\}$ in A. The resulting composition is denoted
$A \oplus [B_1, p_1] \oplus \ldots \oplus [B_n, p_n]$.

Note that this composition is not associative, since $(A \oplus [B]) \oplus [B']$ may have B' riding on some part of A that is unrelated to B, or B' may ride on B.

<u>Definition 2.12</u>. A <u>shell composition</u> is a string that is obtained from the <u>shell string</u>  by adding riders of the same form. (The riders may have riders.)

<u>Definition 2.13</u>. A string is <u>atomic</u> if it is irreducible and has no riders. For example, the trefoil string is atomic, as is the shell string.

In an atomic string A one of the following two forms

       curl   shell

will always be obtained upon removing all interior edges (retaining small interior arcs where an input or output line crosses into a bounded region). Call the first form a <u>curl</u> and the second a <u>shell</u>.

Definition 2.14. We shall define the boundary of a string. It will be a graph ∂A composed of curls and shells, defined inductively as follows:

1. If A is atomic, then ∂A is the single curl or shell obtained (as above) by deleting interior edges.
2. ∂(A ⊕ B) = ∂(A) ⊕ ∂(B)
3. ∂(A ⊕ [B,p]) = ∂(A) ⊕ [∂B,p] when p belongs to ∂A.

   ∂(A ⊕ [B,p]) = ∂(A) when p does not belong to ∂A. (The small interior arcs of the boundary do not contain composition points p.)
4. ∂(⟶) = ⟶ .

Example.

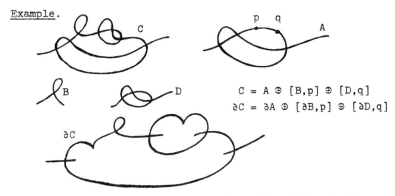

C = A ⊕ [B,p] ⊕ [D,q]
∂C = ∂A ⊕ [∂B,p] ⊕ [∂D,q]

Remark. Since the decomposition of a string into atomic strings is unique, the string boundary is well defined.

With this definition, a composition of curls is its own boundary. The reason for this choice is related to the fact that a curl composition has only one Jordan trail (hence only one state). Since this state has no available state trans-

positions, it satisfies the Clock Theorem vacuously. Curl **compositions** are the only universes with one state.

In order to give a procedure of constructing extremal **states** we must consider the collection $\hat{\mathcal{U}}$ of string universes **with sites**. This contains the ordinary strings $\mathcal{U}$, and it also contains the trails $\mathcal{T}$. (I shall use the same notation for string trails and Jordan trails with sites.) We shall now define a <u>derivative</u> $D : \mathcal{U} \longrightarrow \hat{\mathcal{U}}$. By applying the derivative iteratively to a given string we arrive at a shell composition with sites. This shell composition is then used to construct extremal states.

<u>Definition</u> 2.15. We first define the derivative for atomic strings. In the case of a curl, $D(\,\reflectbox{$\mathcal{C}$}\,) = \underline{\ \Omega\ }$. If A is atomic with a shell boundary, let DA denote the result of splitting the vertices on boundary edges of A so that ∂A injects into DA. This uniquely specifies DA for atoms. Note that DA is a string with sites, and that DA - ∂A = IA is also a string.

<u>Example</u>.

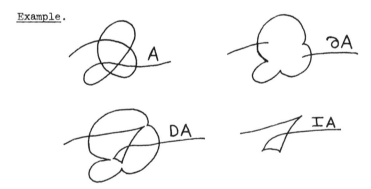

Having defined the derivative for the atomic strings we obtain it generally via:

1. $D(A \oplus B) = D(A) \oplus D(B)$
2. $D(A \oplus [B]) = D(A) \oplus [D(B)]$
3. $D(\longrightarrow) = \longrightarrow\!\!\!\!\succ$

<u>Higher</u> <u>derivatives</u> $D^n : \mathcal{U} \longrightarrow \hat{\mathcal{U}}$ $n = 1, 2, \ldots$ are defined as follows. Let $\mathcal{U}_*$ stand for <u>strings with sites, regarded as strings</u>. That is, $\hat{\mathcal{U}}$ and $\mathcal{U}_*$ have the same information but matters of atomicity and irreducibility are referred soley to the underlying string structure. Let $F : \hat{\mathcal{U}} \longrightarrow \mathcal{U}_*$ be the mapping that associates to a string with sites its corresponding string, and $F^{-1} : \mathcal{U}_* \longrightarrow \hat{\mathcal{U}}$ the inverse map. With this understanding, we have an immediate extension of the derivative to $D : \hat{\mathcal{U}} \longrightarrow \hat{\mathcal{U}}$ via the formula $D(X) = F^{-1}DF(X)$ with the right hand D our original derivative on strings. The higher derivatives now exist via iteration of this operation.

<u>Lemma</u> 2.16. Let A be a string. Then there exists a positive integer N such that $DD^N A = D^N A$. Let $\hat{D}A$ denote this string with sites, and call it the <u>dissection of</u> $\underline{A}$. Then the dissection of a string A is a shell composition with sites.

<u>Proof</u>. This follows easily from the definitions.

In a composition of shells the vertices on each shell can be split in either clockwise or counterclockwise fashion, as illustrated below:

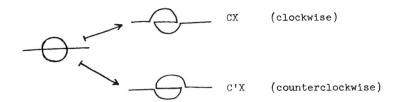

If Y is a composition of shells, we let CY denote the result of performing a clockwise split on each shell (similarly for C'Y).

Theorem 2.17. Let A be a string. Let $KA = C\hat{D}A$ and $K'A = C'\hat{D}A$. Then the trails KA and K'A correspond to clocked and counterclocked states of A respectively.

This theorem completes our statement of the algorithm for constructing extremal states. In order to prove it we shall analyze the sort of sites that may be added to a shell composition to obtain a string with sites of the form $\hat{D}A$. By specifying allowed sites the proof will emerge. Figure 7 illustrates this algorithm as it applies to the example just prior to Definition 2.9.

Allowed and Forbidden Interactions

It is often convenient, when viewing a string with sites, to see it decomposed into the same atoms as the pure string obtained by closing all the sites. To this end we define the map $J : \hat{\mathcal{U}} \longrightarrow \mathcal{U}$ that closes all the sites. A string with sites will be said to be J-atomic or J-irreducible when

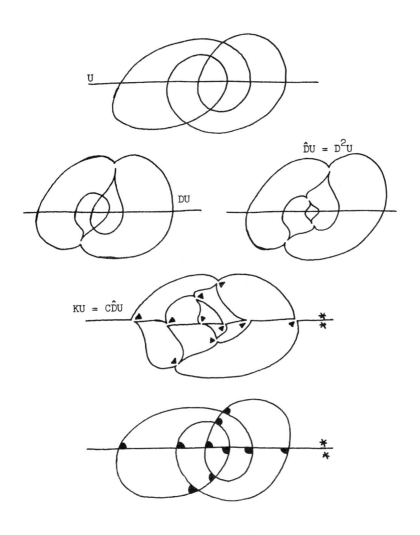

Deriving a Clocked State

Figure 7

its image under J is atomic or irreducible. A pure string (without sites) divides the plane into regions (with two unbounded regions by our conventions). By these same conventions a trail T divides the plane into two regions, but the regions for the corresponding string, $J(T)$, are apparent from the diagram, since they are bounded by edges and sites. The sites correspond to doorways in rooms - one can identify the interior of a room even when the doors are open. Therefore we define the **rooms** of a string with sites $X \in \hat{\mathcal{U}}$ to be the regions of the pure string $J(X)$. Since $J(J(X)) = J(X)$, the regions of a pure string are identical with its rooms (and some rooms may have no doors).

Consider a composition of shells Y and a single circle α in Y. Then α is divided into an upper arc α_+ and a lower arc α_-. Here $\alpha = \alpha_+ \cup \alpha_-$, $\alpha_+ \cap \alpha_- = \{p,q\}$ where p and q are the intersection points of α with a line ℓ. Let α_o denote the intersection of ℓ with the interior (bounded) region determined by α. In Y, the lines α_+, α_- and α_o may each have riders. Suppose that α_+ has riders $U_1, U_2, \ldots, U_r$; α_- has riders $L_1, \ldots, L_s$; α_o has riders $M_1, \ldots, M_t$. Each rider is itself a shell composition, and the riders are all disjoint except for their connections along α_+, α_- and α_o. Let F_α denote this composition of α and its riders. See Figure 8.

33

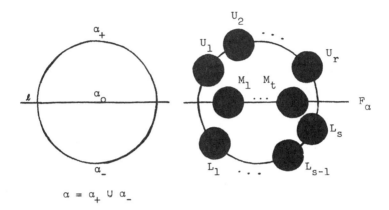

Figure 8

Let $F_{\alpha_{\pm}}$ and F_{α_o} denote $\alpha_{\pm}$ plus its riders, and α_o plus its riders, respectively. Thus F_α = (input and output lines) $\cup F_{\alpha_+} \cup F_{\alpha_-} \cup F_{\alpha_o}$. We wish to specify how sites may be added to F_α, forming a string with sites F'_α, so that $\hat{D}J(F'_\alpha) = F'_\alpha$. A point p of $F'_{\alpha_{\pm}}$ will be said to be a <u>pure boundary point</u> if there exist paths, confined to single rooms of F'_α, from p to the input and output strands of F'_α, and also a path from p to some point of F'_{α_o}. Using this concept, the following interaction rules are adopted:

Interaction rules.

1. Let ───●─── denote a string with sites.
   Then ───●─── may be replaced with either of
   the two forms: ╱‾◠‾╲_◉_╱‾   ‾╲◠╱‾●

2. If F'_α is a string with sites that has been obtained
   from a shell composition of the form F_α, then further
   sites may be added between F'_{α_0} and $F'_{\alpha_\pm}$ as long as
   the point (cusp point) of the site coming from $F'_{\alpha_\pm}$
   is a pure boundary point.

3. Rules 1. and 2. may be applied to any J-rider on a
   given string with sites.

With regard to rule 3., note that a J-rider is simply a substring with sites that has no site interactions with its containing string (hence it corresponds to a decomposable part under J.).

Proposition 2.18. Let SH denote all strings with sites obtained by elaborating shell compositions according to the interaction rules 1., 2., and 3. Let $\mathcal{U}$ denote the collection of site - free strings, and $J : SH \longrightarrow \mathcal{U}$ the mapping that closes all sites.

Then J is a one-to-one correspondence, and $\hat{D} \circ J = 1_{SH}$, $J \circ \hat{D} = 1_\mathcal{U}$. Thus SH is exactly the set of elaborated shell compositions obtained by dissecting strings via $\hat{D}$.

Using the Interaction Rules

Here are a few examples of the application of the interaction rules:

Rule 1:

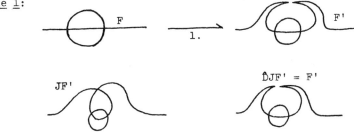

Rule 2:

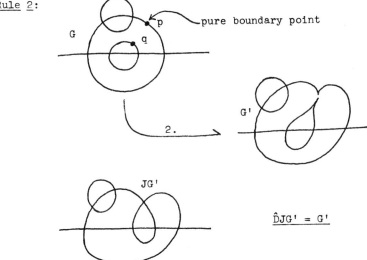

Proof of 2.18. Since $J \circ \hat{D} = 1_{\mathcal{U}}$, it suffices to show that $\hat{D} \circ J = 1_{SH}$. Since this property is preserved under application of the interaction rules, the result follows by induction.

Remark. It is worth observing how things go wrong when the interaction rules are violated. For example, if we add a site to  Then $J(F') = $ and $DJ(F') = $. Thus $DJ(F') \neq F'$ (and F' violates rule 2.).

Before proving Theorem 2.17, we need a lemma about the placement of state markers. Consider an atomic string.

Take its first derivative and start putting in the markers for the clocked state that will result from the algorithm of Theorem 2.17.

Notice that markers at sites between the top of the shell (the shell produced by taking the derivative) and the middle string go to the right in the form   where the +

sign labels the cusp from the top part of the shell. Similarly, the markers for sites between the middle and the bottom part of the shell point to the left. This phenomenon propagates, and we obtain the

Lemma 2.19. Let $X \in SH$ be a shell composition with sites allowed by the interaction rules. Let CX, the result of clocking all shells in C, be decorated with state markers according to the procedure of Theorem 2.4 (states trails correspondence). Let F_α be a shell configuration within X, as depicted in Figure 8. Denote a site with contributing cusps from F_{α_+} and F_{α_0} by the notation  and a site with contributing cusps from F_{α_0} and F_{α_-} by .

Then markers for these sites will appear in CX on the right and left, respectively. That is, they will have the forms

and .

Proof of 2.19. Combine the observation just prior to the statement of this lemma with Proposition 2.18.

Proof of 2.17. Let A be a string. We wish to show that the trail $C\hat{D}A = KA$ corresponds to a clocked state. Since it is easy to see that CX is clocked whenever X is a pure shell composition (without sites), it will suffice (by 2.18) to show that if X' is obtained from $X \in SH$ via the interaction rules, and CX is clocked, then CX' is also clocked. We must show that the addition of an allowed site cannot create a counterclockwise move.

We can limit our considerations to interactions on a form F_α as depicted in Figure 8. Terminology will refer to this figure. Call F_{α_+} the <u>top</u>, F_{α_o} the <u>middle</u>, and F_{α_-} the <u>bottom</u>. Then a new site may be between middle and top, middle and bottom, or it may be a self-interaction of one of these forms. Using the notation of Lemma 2.19, we may assume that all middle - top or middle - bottom sites receive markers in the forms ⤫ and ⤫ . By interaction rule 2., the cusps from the top and from the bottom form parts of the boundary between the interior and the exterior of the curve α. Therefore a counterclockwise move involving either of these markers would necessarily move the marker into the exterior of α. Therefore no interior counterclockwise moves can arise from sites of this type.

Finally, consider the introduction of a top-top, bottom-bottom, or middle-middle site. If this occurs along F_{α_o}, then it will have the form

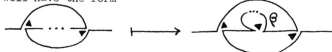

A counterclockwise move then entails an interaction of the form

or

Since the marker at the (x,β) site is on the left and β is part of the middle portion, x must also be part of the middle portion (since top-middle sites have markers on the right). But, this is a forbidden interaction. Each time a self-interaction occurs, a curve or composition β is split off. Further self-interactions must come from β itself. Thus no counterclockwise moves can come from self-interactions of the middle.

Similar arguments apply to the top and bottom. Thus we have shown that CF admits only clockwise internal moves. Since any composition in SH can be decomposed into forms of this type, we have shown that KA does not admit any counterclockwise moves. It remains to prove the existence of clockwise moves in KA.

As we observed after Lemma 2.16, a composition of curls has no available moves. This may occur when the shell composition underlying $\hat{D}A$ is trivial. If, however, this shell composition is non-trivial then KA does admit a clockwise move. Such moves can be located by searching for a <u>deepest shell</u> in $\hat{D}A$. A deepest shell in a shell composition is a shell whose middle, top, and bottom have no riders. Such shells exist since there are a finite number of shells in the composition. It is then easy to see that an elaboration of a deepest shell via the interaction rules will always admit a clockwise move. This completes the proof.

Remark. Figure 9 illustrates some elaborations of riderless shells, and the available clockwise moves.

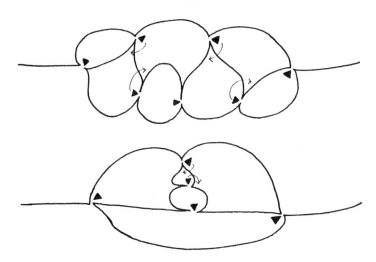

Figure 9

We are now ready to approach the proof of the Clock Theorem. The core of the proof rests on a procedure for going from one state to any other state by a series of transpositions. Once this method is clear, the theorem will follow, and we will be able to show that the extremal states constructed by Theorem 2.17 are unique. In order to create a series of transpositions between two states, we first show that there is a series of <u>exchanges</u> between any two trails. Each exchange then factorizes into a series of clockwise or counterclockwise moves between the corresponding states.

<u>Definition</u> <u>2.20</u>. A trail T' is said to be obtained from a trail T by an <u>exchange</u> if T' is the result of reassembling two sites of T.

Note that upon reassembling a single site, a trail will break up into two components. If these two components interact at another site, then a second reassembly at this site will constitute an exchange. Using string form, the first reassembly produces an extra component that is homeomorphic to a circle. Thus the generic form of the exchange (ignoring the presence of other sites) is as illustrated below.

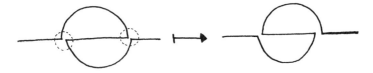

If there are no other sites between the top and middle, or between the middle and bottom, then the exchange is accomplished by a single transposition of the corresponding states. With intervening sites, a series of transpositions can do the job (as we shall prove). Examine Figure 10. Since the generic form of the exchange replaces a clocked form with a counterclocked form (or vice-versa), we shall refer to <u>clockwise and counterclockwise exchanges</u> where a clockwise exchange replaces a clocked form by a counterclocked form. In Figure 10 we see that a clockwise exchange corresponds to a series of clockwise transpositions. This is always the case.

<u>Proposition 2.21</u>. Let T and T' be trails on a universe U. Then there exists a sequence of exchanges taking T to T'.

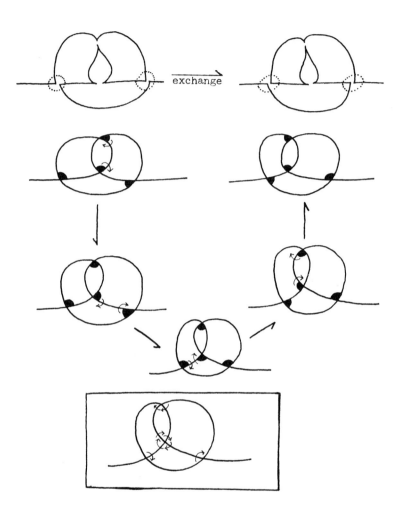

Factorizing an Exchange into Transpositions

Figure 10

Proof of 2.21. Two trails on the same universe must differ at an even number of sites (since one trail can be transformed into the other by reassembling all the sites at which they differ, and an odd number of reassemblies leaves a disconnected form). These sites may be paired off with each other to give the desired set of exchanges.

Proposition 2.22. Let T and T' be trails on a universe U so that T' is obtained from T by one clockwise exchange. Let S and S' be states of U (with the same star placement) corresponding to T and T' respectively. Then S' may be obtained from S by a sequence of clockwise transpositions. Except for the state markers at the exchange sites (which must each turn through 90°), any state marker involved in these transpositions will turn through a total of either 180° or 360°.

Thus clockwise (counterclockwise) exchanges factorize into clockwise (counterclockwise) sequences of transpositions.

Proof of 2.22. We shall prove this result by induction on the number of vertices in the universe U. In order to do this induction it is necessary to state the details of the factorization procedure more precisely. The generic form of the clockwise exchange is that of a clocked shell. As such, it has a top, bottom, and midline as indicated below.

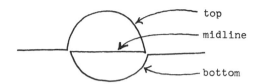

In practice, the top, middle and bottom will all have extra sites. The midline itself is a trail with sites (self-interactions) and cusps (places where the midline has sites with top and bottom). Let λ denote this midline trail with its sites and cusps. Then λ may be written as a sum, $\lambda = \lambda_1 \oplus \lambda_2 \oplus \ldots \oplus \lambda_n$, of J-irreducible trails with sites and cusps, where we extend the notion of J-irreducible and J-atomic (see discussion after 2.17) by taking an isolated cusp (⌒ or ⌣) as J-atomic. Recall that a trail is J-atomic if the string obtained by closing all of its sites is atomic. For the rest of this proof, atomic (irreducible) will always be used in place of the term J-atomic (J-irreducible).

Each λ_K is then a composition of atoms, and the atoms are partially ordered by the relation: $A < B$ whenever A is a rider on B. If $A < B$ and $B < C$ then $A < C$. Note that atoms on different λ_K are unrelated by this partial order.

<u>Induction Hypothesis</u>: The clockwise exchange factors into a sequence of transpositions so that

    1. Only markers at the exchange sites and at sites along the midline trail are utilized.

2. Each exchange site marker rotates clockwise by 90°.

3. For markers that move, those at sites between the midline and the top or bottom will turn a total of 180° clockwise, while those markers at self-interaction sites of the midline will turn a total of 360° clockwise.

4. If A is an atom in the decomposition of the midline, and if A has a cusp rider, then every marker on A will turn.

5. Call an atom <u>involved</u> if all of its markers turn in the factorization. If A is involved, and A < B, then B is involved.

6. Rules 1. through 5. specify exactly the markers (hence the sites) involved in the factorization.

This technical induction hypothesis constitutes an exact statement of Proposition 2.22. Note that it is clearly true for the first few examples, such as that shown in Figure 10. The proposition is also easy to verify for the case where the midline trail has no self-interactions (hence it consists entirely of cusps). We leave this case as an exercise for the reader.

Thus we may assume given an exchange situation on a universe with N vertices, so that the proposition is known for all universes of fewer vertices. If the midline trail has no self-interactions, then we are done by the above remarks. Therefore, it may be assumed that there exists a self-interaction site s on the midline. Remove s via ><{↦) (, s = >< , obtaining a smaller universe U'. Note that all other state markers on U' are the same as those on

U, and that an otherwise identical exchange problem is presented for U'. The induction hypothesis applies to U'.

If s is a site on an uninvolved atom, then no new involvement is created by its removal. Hence, by induction, the factorization for U' extends to a factorization for U that still satisfies the induction hypothesis.

Suppose that s is on an involved atom, and that all the crossings nearest to s remain involved when s is removed. Then, by induction, these nearby markers undergo rotations as described by statements 1. through 5. Upon replacing s, these rotations in U' induce a 360° rotation of the marker at s. The geometry of this induction is illustrated in Figure 11.

Finally, suppose that s is on an involved atom and that when s is removed, one or both of the segments from the site belong to uninvolved atoms in U'. In this case we are presented with a situation in the form

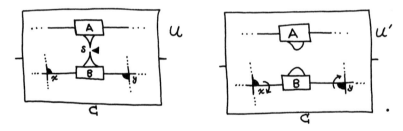

The site s is an interaction site between two atoms A and B. These ride on the larger atom C, which is involved in U. When s is removed, the atoms A and B may no longer be involved. We have illustrated how B could lose involvement:

In U' a direct transposition at x and y is available, and B need not be utilized. Note that by induction, this transposition does occur in the factorization for U' (since the larger atom C is still involved in U'). We are now presented with a small factorization problem of the same type! Namely:

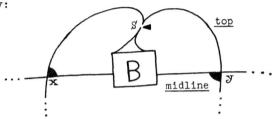

Since this occurs on a smaller universe, the induction hypothesis applies, and we obtain a partial factorization in U where the marker at s turns through 180°. Now apply the same reasoning to A, and obtain the full rotation of 360° for the marker at s.

This completes the induction, and hence the proof of Proposition 2.22.

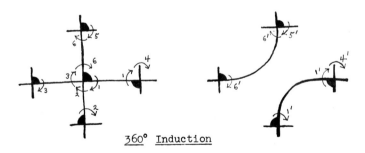

360° Induction

Figure 11

Remark. Here is a concrete example of the last part of the induction argument for 2.22. Let U and U' be as shown below. Thus U' is obtained from U by deleting the site at s (indicated in state form). The atoms A and B in U' are uninvolved, but become part of a larger involved atom C in U. Contained within the larger factorization for U are the small factorizations with A or B and the site s.

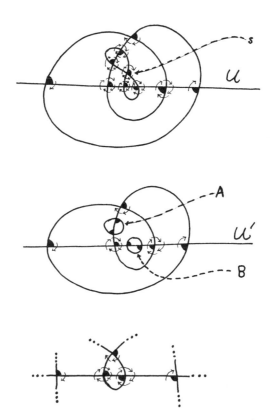

The clocked state is unique.

We are now prepared to show that the clocked state is unique (and that the counterclocked state is unique). In order to do this, consider the form of an atomic trail. If it is a curl form, then no exchanges are available, and it is the only trail on its universe; thus there is nothing to prove. Thus we may assume that the trail T corresponds to an atomic string A that is not a curl. This means that T can be obtained from the schema

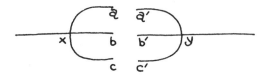

by 1. splitting the vertices at x and y,
2. connecting {a,b,c} and {a',b',c'} so that
   1. and 2. produce a connected curve,
3. adding extra sites except at the input and output lines.

In Figure 12 we have enumerated the three basic possibilities for such an atomic string. Type α illustrates a clocked shell upon which extra sites may be added. We should also include a counterclocked shell α' under this case. In type β the sites x and y have been split in a clocked (or counterclocked) manner, but the pitchfork connections have been made in the one other possible manner. Also illustrated is β_o, the form with the least number of sites that is in this category. The same remarks apply to γ and γ_o.

Any trail in the categories β or γ is obtained by adding extra sites to the prototypes β_0 and γ_0.

Since β_0 and γ_0 admit both clocked and counterclocked exchanges, no trail in these categories can correspond to a clocked or counterclocked state (using Proposition 2.22). Thus a clocked atomic trail is in type α, and a counterclocked atomic trail is in type α'. This determines the outer form of the trail. By repeating this criterion on the resulting midline trail (recursively) we arrive at exactly the description of the clocked state that is summarized in Theorem 2.17.

This completes the proof that the extremal states are unique. Since any state can be transformed into a clocked state by performing successive counterclockwise moves, this shows that any state is connected to the clocked (counterclocked) state by a sequence of transpositions.

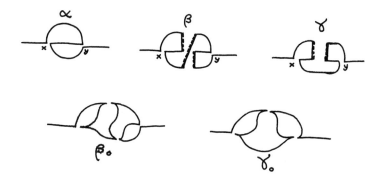

Figure 12

The collection of states is a lattice.

A specific marker in a state can be involved in no more than one transposition at a time. Consequently, it is possible to label all state transpositions from the clocked state on one diagram in the pattern

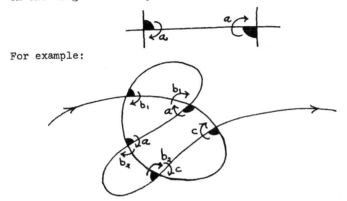

For example:

In this example the moves labelled a, b_1, b_2, c are restricted by the heirarchy

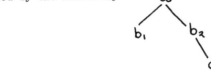

That is, b_1 and b_2 cannot be performed before a, and c must be done after b_2. If we write rs to indicate "do r and then do s", then $rs = sr$ whenever this makes sense. That is, $ab_1 b_2 = ab_2 b_1$ but $ab_2 c \neq acb_2$ since cb_2 is meaningless. Here equality means that the resulting states are identical. Thus, from the heirarchy of operations we

generate the collection of states (Here 1 denotes the clocked state.):

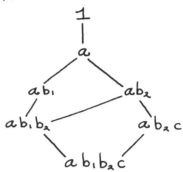

Recall that a <u>lattice</u> is a partially ordered set $\mathcal{S}$ such that every pair of elements X and Y of $\mathcal{S}$ have a well-defined infimum X ∧ Y and a well-defined supremum X ∨ Y. If we indicate states by clockwise move sequences as above, then X ∨ Y is the move sequence corresponding to the intersection of the set of moves for X and Y (Observe that this defines a unique state.). Similarly, X ∧ Y corresponds to the union of the moves for X and Y. With the partial order X < Y whenever there is a series of clockwise moves from Y to X, this gives a lattice structure to the set of states of a string.

The <u>proof</u> <u>of</u> <u>the</u> <u>Clock</u> <u>Theorem</u> is now complete.

3. State Polynomials and the Duality Conjecture

Suppose that labels have been placed in each of the **four** corners of every vertex in a universe U. Let K denote this labelled universe. For purpose of discussion suppose that the labels at the kth vertex, V_k, are

and suppose that the universe has n vertices $(k=1,\ldots,n)$. Define the <u>inner product</u> $\langle K|S\rangle$ between the labelled universe K and a state S of this universe by the formula $\langle K|S\rangle = \sigma(S)V_1(S)V_2(S)\ldots V_n(S)$ where $V_k(S)$ is the label touched by the state marker of S at this vertex, when the state and the labelling forms are superimposed. Thus $V_k(S) = B_k$, W_k, U_k, or D_k according to the position of the state marker of S at the given vertex. $\sigma(S)$ is the sign of the state S.

We regard $\langle K|S\rangle$ as an element of the polynomial ring R whose generators are the collection of labels of K (integer coefficients). The state polynomial for a labelled universe K is then defined by the formula

$$\langle K|\mathcal{S}\rangle = \Sigma_{S\in\mathcal{S}} \langle K|S\rangle.$$

The state polynomial, $\langle K|\mathcal{S}\rangle$, is the determinant of an appropriate matrix: Let $R_1, R_2, \ldots, R_n, R_{n+1}, R_{n+2}$ be an ordering of the regions of U so that if $\mathcal{S}$ is the set of states of U with stars in R_{n+1} and R_{n+2}, then this ordering satisfies Definition 2.7, giving a permutation assignment for $\mathcal{E}$. Define the (generalized) **Alexander matrix** (see [1]) $A(K) = (A_{ij})$ of a labelled universe K to be the $n \times (n+2)$ matrix with columns corresponding to the ordered set of regions and rows to the ordered set of vertices, and entries as indicated below.

$$A_{ij} = \begin{cases} 0 \text{ if } R_j \text{ does not touch } V_i. \\ B_i, W_i, U_i \text{ or } D_i \text{ if } R_j \text{ touches} \\ V_i \text{ in the corner corresponding to} \\ \text{this label. } B_i + W_i \text{ or } U_i + D_i \\ \text{if } R_j \text{ touches two corners at the} \\ \text{vertex } V_i. \end{cases}$$

Let $A(K,\mathcal{E})$ (the reduced Alexander matrix) denote the $n \times n$ matrix obtained from $A(K)$ by striking out the columns R_{n+1} and R_{n+2}.

Proposition 3.1. The state polynomial is the determinant of the reduced Alexander matrix. Thus

$$\langle K|\mathcal{E}\rangle = \text{Det } A(K,\mathcal{E}).$$

Proof. The terms $\langle K|S\rangle = \sigma(S)V_1(S)\ldots V_n(S)$ of the state polynomial correspond exactly to the non-zero terms in the expansion of the determinant of $A(K,\mathcal{E})$. Since (Proposition 2.8) $\sigma(S) = \text{sgn}(P(S))$ for this permutation assignment $P : \mathcal{E} \longrightarrow S(n)$, we see that this proposition follows directly from the definition of the determinant

$$\text{Det } A(K,\mathcal{E}) = \sum_{p \in S(n)} \text{sgn}(p) \, A_{p_1 1} A_{p_2 2} \ldots A_{p_n n}.$$

The state polynomial will be specialized to various sorts of labelling. The reader should compare this formal development with the discussion in the introduction.

The first labelling will be the <u>standard label</u>

 which we shall abbreviate as simply .

A blank corner indicates the label 1.

<u>Proposition 3.2</u>. Let K be the standard labelling of an oriented universe U (i.e., each vertex receives a standard label). Let $\mathcal{S}$ be the set of states for U with a given choice of adjacent fixed stars. Then for a state S in $\mathcal{S}$, $\langle K|S\rangle = (-1)^{b(S)} B^{b(S)} W^{w(S)}$ where $b(S)$ denotes the number of black holes in S, and $w(S)$ denotes the number of white holes in S. Hence $\langle K|\mathcal{S}\rangle = \sum_{b,w} (-1)^b N(b,w,\mathcal{S}) B^b W^w$ where $N(b,w,\mathcal{S})$ is the number of states in $\mathcal{S}$ with b black holes and w white holes.

<u>Proof</u>. This follows directly from definitions.

<u>Remark</u>. Propositions 3.1 and 3.2 taken together show that the states of a universe may be enumerated by taking the determinant of a matrix. This is an analog of the well-known Matrix - Tree Theorem (see [9]). The Matrix - Tree Theorem enumerates rooted trees in a graph. In fact, the states of a universe are in one-to-one correspondence with rooted trees in a graph associated with the universe. This graph, G(U), is obtained as follows: Checkerboard color the regions of U with the colors black and white (say that the unbounded region receives black). The vertices of G(U) are

in one-to-one correspondence with the white regions of U. Two vertices in G(U) are connected by an edge in G(U) whenever the corresponding white regions share a crossing. For example:

It is easy to see from the proof of Theorem 2.4 that <u>maximal rooted trees in G(U) are in one-one correspondence with the set of states of U for a given choice of fixed stars</u>. To see this result, let G'(U) denote the (dual) graph obtained by the same construction on the black regions. Then a maximal rooted tree on either of the graphs G(U) or G'(U) uniquely determines a tree on the other once the roots are specified--together the two trees exhaust the set of crossings in U. A pair of trees specifies a state as in Figure 4.

The Matrix - Tree Theorem has been applied in knot theory via the graphs G(U) and G'(U) (see [8], [9], [12], [33]). To the best of my knowledge, the connection between trees in G(U) and the classical Alexander matrix has not been noted before.

The main result in this section is

<u>Theorem 3.3</u>. Let K be a standard labelling for an oriented universe U. Let δ and δ' be two collections of states for different choices of adjacent fixed stars. Then $\langle K|\delta \rangle = \langle K|\delta' \rangle$. Hence, for each pair of non-negative integers

(r,s), $N(r,s,\delta) = N(r,s,\delta') = N(r,s)$. The number of states with r black holes and s white holes is independent of the choice of fixed stars.

At this point it is worth stating the

Duality Conjecture: Let $N(r,s)$ be defined as in Theorem 3.2., then $N(r,s) = N(s,r)$ for all non-negative integers r, s.

There is good computational evidence for this conjecture. We shall compute one example at the end of the section.

In order to prove Theorem 3.3 we need an indexing of regions due to Alexander ([1]). Each region is assigned an integer index so that adjacent regions have indices differing by one. The increase or decrease of index from region to region depends upon the orientation of the intervening boundary as in the schema below.

Lemma 3.4. Every universe has an Alexander indexing.

Proof. By splitting each vertex in U according to the pattern ⤨ ↦ ⤨ , we obtain a disjoint collection of Jordan curves in the plane, the Seifert circles for U, ([12], [36]). Regions that coalesce under this splitting have the same Alexander index. Hence it suffices to index the regions of a universe that consists of a disjoint collection of oriented Jordan curves. This is clear. (See Figure 13.)

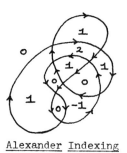
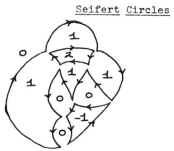

Seifert Circles

Alexander Indexing

Figure 13

We shall use the Alexander indexing to obtain linear dependence relations among the columns of the Alexander matrix. The strategy is as follows: Combinations that sum to zero at each vertex imply global combinations that sum to zero.

More specifically, since the columns of the Alexander matrix are in one-one correspondence with the regions of the universe (see the definition prior to 3.1), we may speak of the columns of index p for an indexed universe. Let C_p denote the sum of the columns of index p. Then, by examining the form of a single corssing, we see that if x satisfies the quadratic equation $x^2 + x(B + W) + 1 = 0$, then $x^{p+2} + x^{p+1}B + x^{p+1}W + x^p = 0$ for any integer p, and we have the global relation $\sum_p x^p C_p = 0$. See Figure 14.

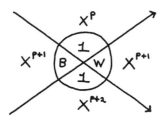

Figure 14

Proof of Theorem 3.3. Let $\rho = B + W$. Then the roots of $x^2 + \rho x + 1 = 0$, denoted α and $\bar{\alpha}$, satisfy the equations $\alpha\bar{\alpha} = 1$ and $\alpha + \bar{\alpha} = -\rho$. Thus, with notation as above, we have $\Sigma_p \alpha^p c_p = 0$ and $\Sigma_p \bar{\alpha}^p c_p = 0$. Hence, letting $[a] = a - a^{-1}$, we have

$$(*) \quad \Sigma_{p \neq k} [\alpha^{p-k}] c_p = 0$$

for any index k (multiply both sums by $\bar{\alpha}^k$ and subtract). Let K be the standard labelling of U, and let $A = A(K)$ be the Alexander matrix. Let $A(k,s)$ be a matrix obtained from $A(K)$ by deleting one column of index k and one column of index s. Let $F(k,s) = \text{Det } A(k,s)$. We may suppose that regions R_{n+1} and R_{n+2} have indices 1 and 0 respectively, and that $A(1,0)$ is obtained by deleting these two columns. Hence, by 3.1, $F(1,0) = \langle K | \delta \rangle$. For $s \neq k \neq r$, the relation $(*)$ yields $[\alpha^{k-r}] c_r = \Sigma_{p \neq k, r} [\alpha^{p-k}] c_p$ and therefore

$(**)$ $[\alpha^{k-r}] F(k,s) = [\alpha^{k-s}] F(k,r)$. To see this identity bring the scalar $[\alpha^{k-r}]$ into the determinant $F(k,s)$ by multiplying a column of index r. Use $(*)$ in the form given above to replace the resulting column by a sum involving all columns of index not equal to k or r. By linearity this determinant can see only a specific column of index s in this sum. Thus the column of index r is effectively deleted and replaced by a column of index s multiplied by $[\alpha^{k-s}]$. Thus proves (**). From this we easily obtain that $[\alpha^{r-t}]F(k,s) \equiv [\alpha^{k-s}]F(r,t)$ whenever $k \neq s$ and $r \neq t$. (Apply (**) twice, once for (k,s,r) and once for (s,r,t).) The symbol $\equiv$ denotes equality up to sign.

In particular, if $k - s = 1$ and $r - t = 1$, then $F(k,s) \equiv F(r,t)$. Since $\langle K|\mathcal{S}'\rangle \equiv \text{Det } A(r,t)$ for some choice of columns so that $r - t = 1$, we conclude that $\langle K|\mathcal{S}\rangle = \langle K|\mathcal{S}'\rangle$ since the sign of a monomial coefficient $B^b W^w$ in either of these polynomials is determined intrinsically as $(-1)^b$. This completes the proof of the theorem.

Remark. We have crucially used the identity between the intrinsically defined state signs and the signs of the permutations that may be associated with the states. This, in turn, depends upon the Clock Theorem. Hence this proof used everything we have built up to this point.

Digression on Star-Separated States

We use the notation of the proof of 3.3. The relationship $[\alpha^{r-t}]F(k,s) \equiv [\alpha^{k-s}]F(r,t)$ gives information about states with widely spaced stars. Let F denote $F(1,0)$, the standard state polynomial, and let F_d denote $F(k,s)$ where $d = k - s$ is the index spacing of the stars. Then

$[\alpha]F_d \equiv [\alpha^d]F$ and hence $F_d \equiv ([\alpha^d]/[\alpha])F$.

Let $\gamma_d = [\alpha^d]/[\alpha]$ where α and $\bar{\alpha}$ are the roots of the quadratic $x^2 + x\rho + 1 = 0$ and $\rho = B + W$.

<u>Proposition 3.5</u>. γ_d satisfies the recursion $\gamma_1 = 1$, $\gamma_2 = -\rho$
$$\gamma_{d+1} = -\rho\gamma_d - \gamma_{d-1}$$

<u>Proof</u>. This follows directly from the formulas $\alpha\bar{\alpha} = 1$, $\alpha + \bar{\alpha} = -\rho$, $\gamma_d = [\alpha^d]/[\alpha] = (\alpha^d - \bar{\alpha}^d)/(\alpha - \bar{\alpha})$.

Thus we have the table

$$\gamma_1 = 1, \quad \gamma_2 = -\rho$$
$$\gamma_3 = \rho^2 - 1$$
$$\gamma_4 = -\rho^3 + 2\rho$$
$$\gamma_5 = \rho^4 - 3\rho^2 + 1$$
$$\gamma_6 = -\rho^5 + 4\rho^3 - 3\rho$$
$$\gamma_7 = \rho^6 - 5\rho^4 + 6\rho^2 - 1$$
$$\ldots$$

The sum of the absolute values of the coefficients of γ_k is the kth Fibonacci number (The Fibonacci numbers are the members of the sequence $1, 1, 2, 3, 5, 8, \ldots, f_k, f_{k+1}, f_k + f_{k+1}, \ldots$). The coefficients themselves are binomial coefficients.

Consider the simplest case of this phenomenon. Each universe in the sequence shown in Figure 15 has $F = 1$ since it has only one state for adjacent fixed stars. If we give the universe U_d its maximal star placement of d, and let

$F_d = F_d(U_d)$ with this placement, then $F_d = \gamma_d F = \gamma_d$. Since there is no cancellation of terms in γ_d when expanded as a polynomial in B and W, γ_d actually lists the states for this star placement. For example, $\gamma_3 = (B + W)^2 - 1 = B^2 + W^2 + 2BW - 1$. Thus we expect U_3 to have five states, one with two black holes, one with two white holes, two with a black hole and a white hole, and one state with no black or white holes. This is born out by the enumeration in Figure 16.

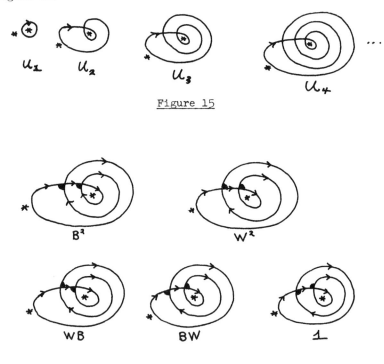

Figure 15

Figure 16

To obtain a purely geometric view of the same situation let $\Gamma_d = \Sigma_{S \in \mathcal{E}_d} B^{b(s)} W^{w(s)}$ where $\mathcal{E}_d$ denotes the set of states of U_d with farthest star placement as in Figure 15. Show, using only the geometry and an induction argument, that $\Gamma_1 = 1$, $\Gamma_2 = \rho$ and $\Gamma_{d+1} = \rho \Gamma_d + \Gamma_{d-1}$ where $\rho = B + W$.

In the general case the polynomial F_d will not explicitly list all the states. This is due to cancellation of terms in the determinant expansion. Also, signs for star-separated states cannot be computed simply from a black hole count. We leave it as an adventure for the reader to look at further properties of states with non-adjacent stars!

The Duality Conjecture

We end this section by computing an example that illustrates the Duality Conjecture. The example is based on the universe U shown in Figure 18. Here we have drawn the clocked state and illustrated the operation heirarchy (see the example at the end of section 2.). Also listed is the effect that each operation has on the black-white hole count: For example, BW^{-1} next to the operation b_2 indicates that this operation adds a black hole and subtracts a white hole. Figure 19 shows the state lattice generated from the hierarchy. Finally, Figure 20 shows the black-white hole counts computed from this lattice. The Duality Conjecture is confirmed for U.

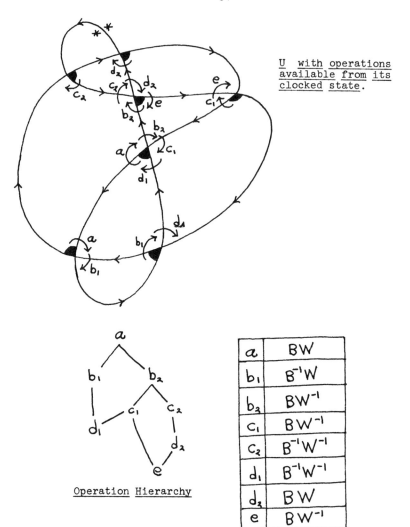

Figure 18

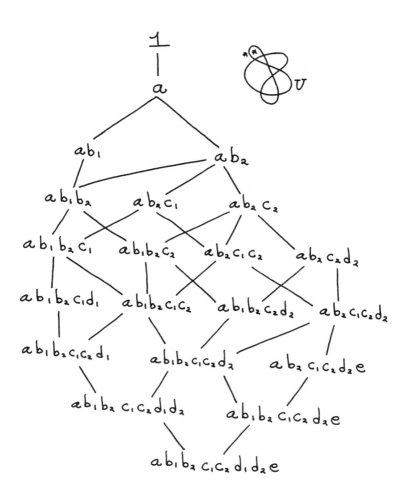

<u>State lattice L(U) generated from the operation heirarchy</u>.

<u>Figure 19</u>

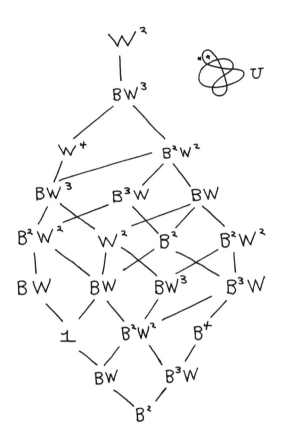

Black Hole - White Hole Classification of L(U)
N(r,s) = Number of states in form $B^r W^s$
Duality Conjecture: N(r,s) = N(s,r).

Figure 20

4. Knots and Links

We generalize the state polynomial from universes to knots and links. The key to this ascent into a third dimension is entailed in the code for crossings:

$$\diagup\!\!\!\!\diagdown \;\equiv\; B\diagdown\!\!\!\!\diagup W \qquad \text{(standard)}$$

$$\diagup\!\!\!\!\diagdown \;\equiv\; W\diagdown\!\!\!\!\diagup B \qquad \text{(reverse)}$$

A knot or link (diagram) is an oriented universe with standard or reverse labels at each crossing.

Given a diagram K with underlying universe U, and a state S of U, we have defined (section 3) the inner product $\langle K|S \rangle \in Z[B,W]$. In this case $\langle K|S \rangle$ is given by the formula $\langle K|S \rangle = \sigma(S) B^x W^y$ where x denotes the number of coincidences of B-labels with state markers of S that occur when the diagrams for K and S are superimposed. Similarly, y denotes the number of W-coincidences.

It is instructive to compute the state polynomial by superposition, and also via the Alexander matrix. Such a computation is shown for the trefoil knot in Figure 21.

<u>Theorem 4.1.</u> Let K be a knot or link with underlying universe U. Let $\mathcal{S}$ be the set of states of U for a given choice of adjacent stars. Let $\langle K \rangle = \langle K|\mathcal{S} \rangle$. Then the polynomial $\langle K \rangle$ is independent of the choice of stars. Furthermore, if K, $\overline{K}$, L are links that differ at the site of <u>one</u>

crossing as indicated below

then $\langle K \rangle - \langle \overline{K} \rangle = (W-B)\langle L \rangle$.

In other words,

<u>Proof</u>. It may happen that the universe U corresponding to K and $\overline{K}$ is connected, while the universe U' corresponding to L is disconnected. In this case K, $\overline{K}$, and L must have the form indicated below

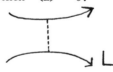

This observation justifies setting $\langle L \rangle = 0$ when the underlying universe is disconnected. For it is not possible to have adjacent stars in a state of the form . Hence, when U' is disconnected then all states of U will have either up or down markers at the vertex in question. Therefore $\langle K \rangle = \langle \overline{K} \rangle$ and so $\langle K \rangle - \langle \overline{K} \rangle = 0 = \langle L \rangle$.

We reformulate this observation by stating: If the vertical dotted line between the two strands of L does not cleave a region of U', then $\langle L \rangle = 0$.

Now assume that this dotted line does cut a region. Letting $\mathcal{S}'$ denote the states of U', we see that $\mathcal{S}' = \mathcal{L} \cup \mathcal{R}$ where $\mathcal{L}$ denotes the states of $\mathcal{S}'$ with a marker to the left of the dotted line, and $\mathcal{R}$ the states with marker to the right of the dotted line. (If the region in U' between the two strands is occupied by a star, let the star reside on one side or the other of the dotted line. In this case either $\mathcal{L}$ or $\mathcal{R}$ is empty.) By closing the site at the dotted line, and adding a marker on the unoccupied side, we obtain one-to-one correspondences of states of U' with a subset of states of U. $\mathcal{L}$ is in one-one correspondence with $\mathcal{W}$, and $\mathcal{R}$ with $\mathcal{B}$ where $\mathcal{W}$ and $\mathcal{B}$ are the states of U with a white hole or a black hole respectively at this crossing

$$\mathcal{L} \longleftrightarrow \mathcal{W} :$$

$$\mathcal{R} \longleftrightarrow \mathcal{B} :$$

Thus $\langle K|\mathcal{W}\rangle = W\langle L|\mathcal{L}\rangle$ $\langle \overline{K}|\mathcal{W}\rangle = B\langle L|\mathcal{L}\rangle$

$\langle K|\mathcal{B}\rangle = -B\langle L|\mathcal{R}\rangle$ $\langle \overline{K}|\mathcal{B}\rangle = -W\langle L|\mathcal{R}\rangle$.

(The black holes contribute the minus signs via the formula for the sign of a state $S : \sigma(S) = (-1)^{b(S)}$.) Hence $\langle K|\mathcal{W} \cup \mathcal{B}\rangle - \langle \overline{K}|\mathcal{W} \cup \mathcal{B}\rangle = (W - B) \langle L|\mathcal{L} \cup \mathcal{R}\rangle$. Since $\langle K\rangle - \langle \overline{K}\rangle$ cannot see up - states or down - states, $\langle K\rangle - \langle \overline{K}\rangle = \langle K|\mathcal{W} \cup \mathcal{B}\rangle - \langle \overline{K}|\mathcal{W} \cup \mathcal{B}\rangle \Rightarrow \langle K\rangle - \langle \overline{K}\rangle = (W-B)\langle L\rangle$. This completes the proof of the exchange identity.

We use the exchange identity to prove independence of star

placement. Since L has fewer crossings than K or $\bar{K}$, we may assume by induction that ⟨L⟩ is independent of the choice of stars. Hence, by the exchange identity, ⟨$\bar{K}$⟩ is independent if ⟨K⟩ is independent. Since any two links with the same underlying universe can be connected by a sequence of crossing exchanges, it suffices to produce one K so that ⟨K⟩ is independent of the star choice. This is provided by Theorem 3.3 for K the standard labelling. The proof of 4.1 is now complete.

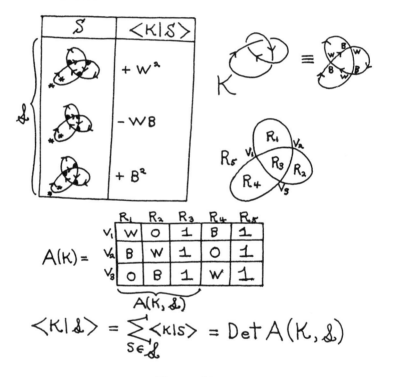

Figure 21

Topological Invariance

We now investigate how $\langle K \rangle$ changes under elementary deformations of the link diagram. The three basic deformations (the Reidemeister moves [33]) are as shown in Figure 22.

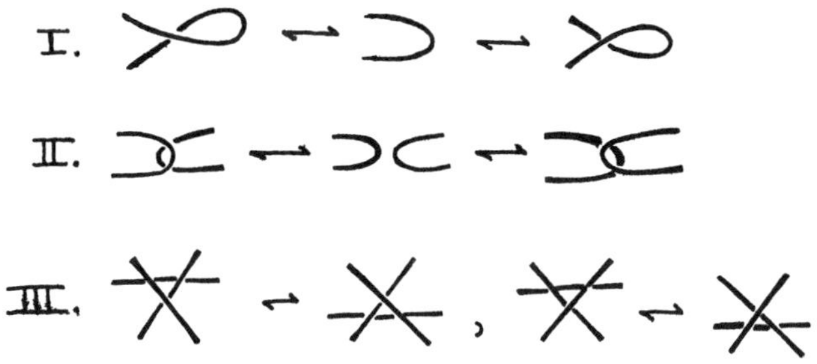

I. ⤴ ↔ ⊃ ↔ ⤴

II. ⤬ ↔ ⊃⊂ ↔ ⤬

III. ✳ ↔ ✳ , ✳ ↔ ✳

<u>Elementary Moves</u>
Figure 22

Figure 22 shows representative situations for each move. The diagram must have the local forms as shown in this figure; the move is performed without disturbing the rest of the diagram.

<u>Definition</u> 4.2. Two link diagrams K and K' are <u>equivalent</u> (K ~ K') if there is a sequence of elementary moves carrying K to K'. (Two diagrams whose underlying universes are isomorphic as planar maps are regarded as identical. Thus topological deformations of the underlying universe are placed in the background of this discussion. Such deformations have no effect on the set of states or on the state polynomials.)

Remark. Equivalence of link diagrams corresponds to ambient isotopy of the corresponding links in three-dimensional space (see [33]).

We shall see that $\langle K \rangle$ becomes an invariant of the equivalence class of K if we set $WB = 1$. In order to formalize this collapse, note that if I is the ideal in $Z[B,W]$ generated by $BW - 1$, then $Z[B,W]/I$ is isomorphic to the ring $Z[B,B^{-1}]$. Let $\psi : Z[B,W] \longrightarrow Z[B,W]/I$ denote this quotient homomorphism. We then define the <u>Alexander-Conway Polynomial</u> <u>of a link</u> K by the formula $\nabla_K = \psi \langle K \rangle$. We shall see that ∇_K is a polynomial in $z = W - B$.

<u>Theorem</u> 4.3. Let ∇_K denote $\langle K \rangle$ when $BW = 1$ (as above), and let $z = W - B$. Then

1. $K \sim K' \Rightarrow \nabla_K = \nabla_{K'}$
2. $K \sim 0 \Rightarrow \nabla_K = 1$ (0 denotes the trivial knot)
3. $\nabla_K - \nabla_{\overline{K}} = z\nabla_L$ when K, $\overline{K}$, and L are related as in Theorem 4.1.

Furthermore, these three properties suffice to calculate ∇_K without any reference to its definition as a state polynomial.

Remark. In section 5 we will take 1., 2., and 3. above as <u>axioms</u>, and show that they define a unique invariant of knots and links. Consequently, the last part of 4.3 (calculation from the axioms) is deferred to the next section.

Proof of 4.3. Since <K> is independent of the star placement (by 4.1), we shall choose those star locations that are most convenient in the calculations to follow.

For elementary moves of type I. we assume that the curl is star-free and hence must contain a marker. Thus there is one-to-one correspondence of states:

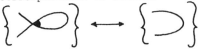

Since the marker in the curl is either an up or down type (depending upon the orientation of this string), this correspondence of states shows that <K> is invariant under moves of type I.

For moves of type II. we distinguish two cases:

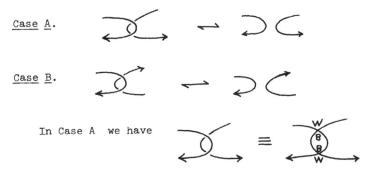

Case A.

Case B.

In Case A we have

Choosing star location as shown below, the states fall into the types S_0, S_1, S_2:

We remark that the regions X and Y above and below these diagrams may be assumed distinct. For if X and Y connote the same region, the situation is of type

K: ▨ (X=Y).

Hence there are no states of type S_0 when X = Y. Since S_1 and S_2 contribute equally with opposite sign, we conclude that

⟨▨⟩ = 0 = ⟨▨⟩

Now return to Case A when X ≠ Y. Then there is a one-to-one correspondence of states

{ ⟩⟨ } ↔ { ⟩ ⟨ }.

It follows that

⟨⟩⟨⟩ = BW⟨⟩ ⟨⟩.

In Case B we have the coding

Again, states of type S_1 and S_2 cancel each other's contribution to the state polynomial. From the code above and the correspondence of states, it then follows that ⟨K⟩ is invariant under the move of type II B. This completes the verification that $\triangledown_K$ is invariant under moves of type I. and II.

Analysis of the cases for the moves of type III. will proceed by listing state types for the star placements

These indicate corresponding regions before and after the move (compare with Figure 22).

For example, suppose we consider a state of the form

Here the ⊗ signs occupy regions that have state markers other than at this triangle. The state depicted above is the only type for the horizontal line in the upper position. Sliding the horizontal line downward, we find three types S_0', S_1', S_2':

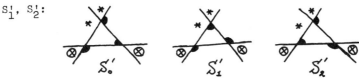

In order to catalog the contributions of these states, suppose that K and K' (differing by a type III move) are given locally by the diagrams

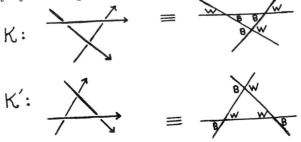

Let α denote the contribution to $\langle K|S\rangle$ from outside the given triangle. Then $\langle K|S\rangle = BW\alpha$ while $\langle K'|S_0'\rangle = -W^2\alpha$, $\langle K'|S_1'\rangle = \alpha$, $\langle K'|S_2'\rangle = W^2\alpha$. Hence $\langle K'|S_0' \cup S_1' \cup S_2'\rangle = \alpha$. This proves that $\psi\langle K|S\rangle = \psi\langle K'|S_0' \cup S_1' \cup S_2'\rangle$. Identical sorts of calculations go through in all the remaining cases (The cases are listed in Figure 23.). This proves the invariance of ∇_K under type III. equivalence.

Since part 3. of this theorem has been proved in Theorem 4.1, this completes the proof of parts 1., 2., and 3. of Theorem 4.3.

<u>Discussion.</u> In Section 5 we shall discuss how to calculate directly from the axioms. Here it is appropriate to record a few examples via state summation:

1. If K is a trefoil knot, then we see from Figure 21 that $\nabla_K = 1 + z^2$, since $\langle K\rangle = W^2 - WB + B^2 = (W - B)^2 + WB$.

2. If L is the link of two unknotted circles of linking number 1, then $\langle L\rangle = W - B$ and $\nabla_L = z$.

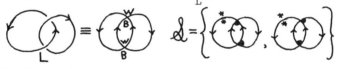

3. If K' is the standard knot corresponding to the universe of Figure 1, then K' is as shown below and it is easy to see from Figure 1 that the state polynomial is $\langle K'\rangle = 2W^2 + 2B^2 - 3WB = 2(W-B)^2 + WB$. Hence $\nabla_{K'} = 1 + 2z^2$.

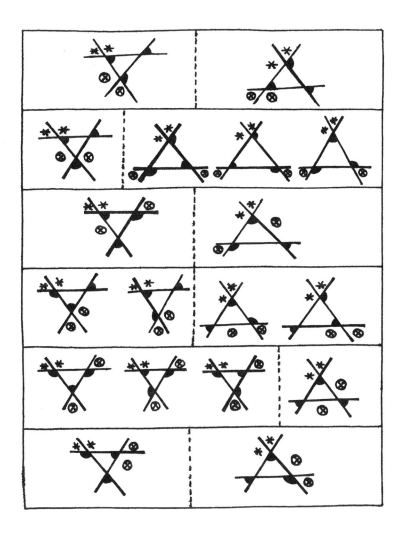

Triangle States

Figure 23

5. Axiomatic Link Calculations

Except for the definition of equivalence of link (diagrams) via elementary moves (See Figure 22.), this section is self-contained. We discuss, with many examples, the subject of calculating the Alexander-Conway polynomial, ∇_K, solely from the axioms. Henceforth, this polynomial will be referred to as the <u>Conway Polynomial</u>. It's axioms are a partial list of properties given by Conway in [5] (with a slightly different normalization). Conway's paper discusses calculations of the kind we are about to do; it does not give models for the axioms or discuss consistency. Here, of course, we have already <u>proved</u> the axioms in Theorem 4.3. Nevertheless, this section will end with a brief discussion of Ball and Mehtah's proof that the axioms are self-consistent. They provide their own model!

Axioms for the Conway Polynomial

1. To each oriented knot or link K there is associated a polynomial $\nabla_K(z) \in Z[z]$ such that equivalent (ambient isotopic) links receive identical polynomials.

2. If K is an unknotted circle, then $\nabla_K = 1$.

3. If K, $\overline{K}$, and L are three links that differ at the site of one crossing as indicated below, then $\nabla_K - \nabla_{\overline{K}} = z \nabla_L$.

When links A, B and C are (diagrammatically) related in the form

[diagram: three crossings labeled A, B, C — A shows a negative crossing, B shows a positive crossing, C shows two parallel strands]

we shall write $A = B \oplus C$ and $B = A \ominus C$. Thus, by axiom 3., $\nabla_{A \oplus B} = \nabla_A + z\nabla_B$ and $\nabla_{A \ominus C} = \nabla_A - z\nabla_C$. Regard $\oplus$ and $\ominus$ as non-associative, non-commutative binary operations defined on appropriately related pairs of links. (<u>Warning</u>: These operations should not be confused with the string combinations of section 2. Since we do not deal with strings outside of section 2, I have taken the liberty of using the same notation for these different operations.)

While we work with knots and links via planar diagrams for theoretical purposes, it is worth noting that they can just as well be seen as collections of embedded and oriented circles in three-dimensional space. For example, here is a standard procedure for producing an unknotted circle from a given diagram: <u>Choose a point p on the diagram and draw a knot so that you first draw an over-crossing line at the first encounter with a crossing, undercross at the second encounter, and continue until you return to p</u>.

It is easy to see that this procedure produces an unknot, but quite a task to show it via the elementary moves. See Figure 24 for an example of this unknotting method.

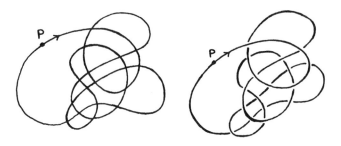

<u>Producing an Unknot</u>

Figure 24

A link L is said to be <u>split</u> if it is equivalent to a union of non-empty sublinks L_1 and L_2 that are contained in disjoint regions of the planar diagram (or in disjoint three - balls in three dimensional space).

<u>Lemma</u> <u>5.1</u>. $\nabla_L = 0$ when L is a split link.

<u>Proof</u>. We may assume that L has a diagram in the form given below:

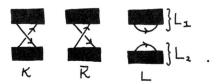

Since K is ambient isotopic to $\bar{K}$ via a 2π rotation, axiom 1. implies that $\nabla_K = \nabla_{\bar{K}}$. By axiom 3., $\nabla_K - \nabla_{\bar{K}} = z\nabla_L$. Hence $0 = z\nabla_L$. Therefore $\nabla_L = 0$. We are now prepared to calculate the trefoil knot.

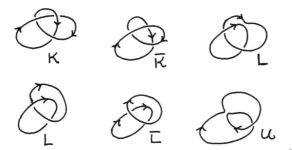

Figure 25

Refer to Figure 25. We have $K = \overline{K} \oplus L$, $L = \overline{L} \oplus U$, where $\overline{K}$ is an unknot and U is an unknot. Since $\overline{L}$ is equivalent to a split link, $\nabla_{\overline{L}} = 0$. Therefore $\nabla_K = 1 + z\nabla_L$, $\nabla_L = z$ and $\therefore \nabla_K = 1 + z^2$.

In this example we can write $K = \overline{K} \oplus (\overline{L} \oplus U)$. This is a <u>skein decomposition</u> of K. The individual parts of a skein decomposition are unknots (O) and unlinks (O O O ... O). We refer to these as the <u>generators</u> of the skein. (This terminology is due to Conway [6].)

<u>Lemma 5.2</u>. Every link L has a skein decomposition into generators. Hence the Conway polynomial may be calculated via the axioms.

<u>Proof</u>. If L is a knot then there exists a sequence of crossing changes (⤫ ⇌ ⤫) that will unknot it. If L is a link, then some sequence of crossing changes will produce a collection of unknotted, unlinked components

Hence there is a sequence $L = L_0, L_1, \ldots, L_n$ such that L_{i+1} is obtained from L_i by switching one crossing, and L_n is a skein generator. Thus we have

$$L = L_0 = L_1 \oplus \epsilon_1 L_1'$$
$$L_1 = L_2 \oplus \epsilon_2 L_2'$$
$$\ldots$$
$$L_{n-1} = L_n \oplus \epsilon_n L_n'$$

where $X \oplus -Y = X \ominus Y$ and $\epsilon_k = \pm 1$. Here L_k' is obtained from L_{k-1} and L_k by splicing out the switched crossing. This gives a skein decomposition of L into L_n and $\{L_1', L_2', \ldots, L_n'\}$. Since the links in this latter set have diagrams with one fewer crossing than the diagram L, we may assume inductively that each has a decomposition into generators. This completes the proof of the lemma.

Remark. Note that the method of proof of this lemma is the way we computed the polynomial for the trefoil knot. Following the notation of the proof, we may conclude that $\nabla_L = \nabla_{L_n} + z \sum_{k=1}^{n} \epsilon_k \nabla_{L_k'}$. Hence $\nabla_L = 1 + zD$ when L is a knot, and $\nabla_L = zD$ when L is a link with more than one component.

A <u>connected sum</u> of links K and K', denoted $K \# K'$, is defined by splicing two strands as shown in Figure 26. This may depend upon the choice of strands.

Figure 26

Proposition 5.3.

1. If $K \# K'$ denotes a connected sum of the links K and K', then $\nabla_{K \# K'} = \nabla_K \nabla_{K'}$.

2. If K^* is the result of reversing all orientations on all strands of K, then $\nabla_{K^*} = \nabla_K$.

3. Let $K^!$ be the mirror image of K (obtained by switching every crossing in K). Then $\nabla_{K^!}(z) = \nabla_K(-z)$.

4. If L is a link with λ components, then $\nabla_L(-z) = (-1)^{\lambda+1} \nabla_L(z)$. Hence $\nabla_{L^!} = (-1)^{\lambda+1} \nabla_L$.

The proof of this proposition will be omitted. All these results follow easily by induction and the fact that the skein is generated by trivial knots and links. Note that part 4. implies that a link with an even number of strands is inequivalent to its mirror image whenever it has a non-zero Conway polynomial.

Examples

1.

$L = \overline{L} \oplus K$, $\nabla_{\overline{L}} = -z$, $\nabla_K = 1 + z^2$

$\therefore \nabla_L = -z + z(1+z^2) = z^3$.

2. Let M be the knot shown below. Then $M = \overline{M} \oplus Q$ where Q is isotopic to two unlinked circles, and $\overline{M}$ is isotopic to $K \# K$ where K is a trefoil knot.

The knots M and $K \# K$ are distinct (as can be verified by other means) but they have the same polynomial.

3. Let K_n $n = 1,2,3,\ldots$ be the sequence of links indicated below. Let $\nabla_n = \nabla_{K_n}$. Then $\nabla_1 = 1$, $\nabla_2 = z$ and $\nabla_{n+1} - \nabla_{n-1} = z\nabla_n$ (by axiom 3).

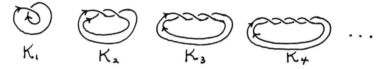

Hence
$$\nabla_1 = 1, \nabla_2 = z, \nabla_3 = z^2 + 1$$
$$\nabla_4 = z^3 + 2z$$
$$\nabla_5 = z^4 + 3z^2 + 1$$
$$\nabla_6 = z^5 + 4z^3 + 3z$$
$$\nabla_7 = z^6 + 5z^4 + 6z^2 + 1$$
$$\nabla_8 = z^7 + 6z^5 + 10z^3 + 4z^2$$

$\ldots$

4. Let K be the standard knot (all crossings of type ⤳) associated with the universe of Figure 1. Then K has diagram

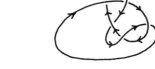

We leave it to the reader to show $\nabla_K = 1 + 2z^2$. This is a good example to compute via the axioms in more than one way. It can also be computed from the lattice of states given in Figure 1.

5. Let K' be the standard knot associated with the universe of Figure 18. Then K' has diagram

and we see that K' is equivalent to K_5 of example 3. Hence $\nabla_{K'} = z^4 + 3z^2 + 1$.

From Figure 20 it is easy to deduce that K' has state polynomial (see section 4)

$$\langle K' \rangle = (W-B)^4 + (WB+2)(W-B)^2 + 1.$$

Hence (via $z = W - B$ and $WB = 1$ under ψ) we see that $\psi\langle K'\rangle = \nabla_{K'}$, as predicted by Theorem 4.3.

6. <u>A Weave</u>: Let $\wedge \mapsto R\wedge$ be the operation indicated in Figure 27. Then, as shown in this Figure, $\overline{R^n \wedge} \sim \overline{\wedge}$.

<u>Claim</u>: $\quad \nabla_{\overline{R^n \wedge}} = \nabla_{\overline{\wedge}} - z^2 \nabla_{\overline{R^{n-1} \wedge}}.$

Proof. Let $X_n = R^n \wedge$. Thus $\overline{X}_n \sim \wedge$.

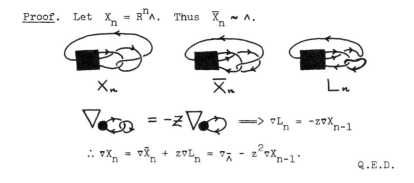

$$\therefore \nabla X_n = \nabla \overline{X}_n + z\nabla L_n = \nabla_{\overline{\wedge}} - z^2 \nabla X_{n-1}.$$

Q.E.D.

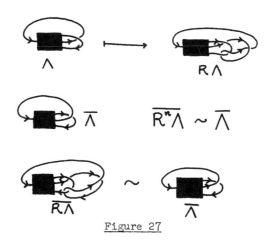

Figure 27

In Figure 27 a knot $\wedge$ is given with unknown tying within the black box, and some particular starting configuration. The transform $R\wedge$ is a knot of the same form. The claim shows that we may recursively calculate the polynomials for these knots. For example, let $\wedge$ be a trefoil knot. Then $\overline{\wedge}$ is an unknot, and we have (letting $f_n = \nabla_{X_n}$) $f_0 = 1 + z^2$, $f_n = 1 - z^2 f_{n-1}$.

$X_0 = \wedge \quad X_1 = R\wedge \quad X_2 = R^2\wedge \quad X_3 = R^3\wedge \quad \ldots$

$$f_0 = 1 + z^2 \qquad\qquad f_\infty = \lim_{n\to\infty} f_n = 1 - z^2 + z^4 - z^6 \pm \ldots$$

$$f_1 = 1 - z^2 - z^4$$

$$f_2 = 1 - z^2 + z^4 + z^6 \qquad f_\infty = 1/(1+z^2).$$

...

It is tempting to try to make sense out of f_∞ by assigning f_∞ as an extended invariant for an infinite knot K_∞ such that $RK_\infty \sim K_\infty$. For a possible formalization of this idea, see [20].

7. Tangle Theory

This example is a quick introduction to Conway's theory of tangles. The material here can be generalized in various ways. See [5] and [15] for more information. A _tangle_ is a piece of a knot diagram with two input strings and two output strings, oriented as in Figure 28. Each input is connected to one output, and there may be any other knotting or linking (without free ends) inside the tangle box. Given tangles A and B, the tangle A + B is defined by connecting inputs to outputs as in Figure 28. Also, there are two ways to form a knot or link from a given tangle A. These are denoted N(A) (numerator) and D(A) (denominator) as in Figure 28.

Given a tangle A, let $A^N = \nabla_{N(A)}$ and let $A^D = \nabla_{D(A)}$. The _fraction_ of the tangle A is then defined by the formula $F(A) = A^N/A^D$.

Theorem 5.4. Let A and B be tangles. Then $F(A+B) = F(A) + F(B)$. That is, $(A+B)^N = A^N B^D + A^D B^N$ and $(A+B)^D = A^D B^D$.

For example, consider the tangles $\underline{0}$ and $\underline{\infty}$:

$\underline{0}$ = ⇄ $\underline{\infty}$ = ⊃⊂

$\underline{0} + \underline{0}$ = ⇄ = $\underline{0}$ $\underline{\infty} + \underline{0} = \underline{\infty}$

$\underline{0} + \underline{\infty}$ = ⊃⊂ = $\underline{\infty}$ $\underline{\infty} + \underline{\infty}$ = ⊃⊂⊃⊂ = $\underline{\infty}_{-1}$

$\therefore F(\underline{0}+\underline{0}) = \frac{0}{1} = \frac{0}{1} + \frac{0}{1} = F(\underline{0}) + F(\underline{0})$

$F(\underline{0}+\underline{\infty}) = F(\underline{\infty}) = \frac{1}{0} = \frac{0}{1} + \frac{1}{0} = F(\underline{0}) + F(\underline{\infty})$

$F(\underline{\infty}+\underline{\infty}) = F(\underline{\infty}_{-1}) = \frac{0}{0} = \frac{1}{0} + \frac{1}{0} = F(\underline{\infty}) + F(\underline{\infty})$.

Note that the identity $0/0 = 1/0 + 1/0$ follows from formal addition of fractions $a/b + c/d = (ad + bc)/bd$. Thus we have verified the theorem for the tangles $\underline{0}$ and $\underline{\infty}$. It will be left as an exercise for the reader to show that this actually suffices to prove the theorem! (Compare with Lemma 5.2.)

As an application of this addition theorem, let L_n denote the link illustrated in Figure 29. It then follows from 5.4 that $\nabla_{L_n} = \text{Num}\left(z + \frac{1}{z} + \frac{1}{z} + \ldots + \frac{1}{z}\right)$ where this formula denotes a continued fraction with n terms.

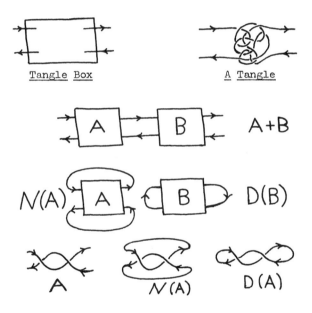

Figure 28

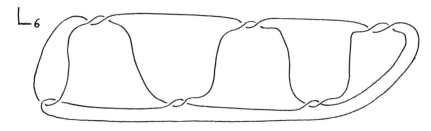

Figure 29

8. The Coefficients

Let $\nabla_K(z) = a_0(K) + a_1(K)z + a_2(K)z^2 + a_3(K)z^3 + \ldots$.
Then the coefficients of the polynomial are themselves invariants of link type. It follows immediately from the axioms that $a_0(K) = \begin{cases} 1 & \text{if } K \text{ is one component} \\ 0 & \text{otherwise.} \end{cases}$

For certain calculations it is useful to adopt a notation to indicate the operations of switching and eliminating a crossing. Accordingly, let

$$E(\overcrossing) = \smoothing = E(\undercrossing)$$
$$S(\overcrossing) = \undercrossing \qquad S(\undercrossing) = \overcrossing ,$$

and let $E_i K$ and $S_i K$ denote the result of applying these operations to the ith crossing of K.

We have implicitly, up until now, assigned indices to the crossings according to the formulas:

$$I(\overcrossing) = +1 \quad , \quad I(\undercrossing) = -1.$$

Let $I_i K$ denote the index of the ith crossing of K. Then the exchange identity of axiom 3 may be expressed by the formula

$$\nabla_K - \nabla_{S_i K} = I_i K z \nabla_{E_i K}.$$

It follows that the coefficients of the Conway polynomial obey the series of identities:

$$a_{n+1}(K) - a_{n+1}(S_i K) = I_i K a_n(E_i K).$$

In particular, it is now easy to see that $a_1(K)$ has an interpretation in terms of linking numbers:

Definition 5.5. Let K be a two-component link. Define the <u>linking number</u>, $Lk(K)$, by the formula $Lk(K) = \frac{1}{2} \Sigma_{i \in T(K)} I_i(K)$. Here the set $T(K)$ denotes the set of crossings of different components of K. Self-crossings do not contribute to the linking number. For example

$$Lk\left(\;\right) = +1.$$

Lemma 5.6. $a_1(K) = \begin{cases} Lk(K) & \text{if } K \text{ has two components,} \\ 0 & \text{otherwise.} \end{cases}$

Proof. This follows at once from the definition of the linking number, and from the exchange identity for a_1.

Remark. It is obvious from its definition that the linking number is invariant under elementary moves.

The problem of direct interpretation for $a_2(K)$ and for the higher coefficients of the Conway polynomial is more mysterious! We shall now give an interesting property of $a_2(K)$ taken modulo-2, and in section 10 we shall show that $a_2(K)$ (mod-2) is the Arf invariant ([34]) of the knot K.

Theorem 5.7. Let K and K' be two knots that are related according to the diagram below:

That is, the diagrams for K and K' are otherwise identical. Then $a_2(K) \equiv a_2(K')$ (modulo-2).

Proof. Label the vertices at the four-fold crossing as indicated below.

Let $X_1 = E_1 K$, $X_2 = E_2 S_1 K$, $X_3 = E_3 S_2 S_1 K$, and $X_4 = E_4 S_3 S_2 S_1 K$. Then a repeated application of axiom 3 implies that
$\nabla_K - \nabla_{K'} = z(\nabla_{X_1} - \nabla_{X_2} + \nabla_{X_3} - \nabla_{X_4})$. In particular,
$a_2(K) - a_2(K') = a_1(X_1) - a_1(X_2) + a_1(X_3) - a_1(X_4)$. This formula reduces the problem of the change of a_2 to a pattern of linking numbers for the four associated links X_1, X_2, X_3, X_4. The rest of the proof is left as an exercise. Note that there are three basic forms of connection for the four-fold crossing to form a knot (one component). They correspond to the schemas:

Remark: Theorem 5.7 also holds for a four-fold crossing switch with reversed orientations on one pair of parallel strands. Thus K and K' may have the local forms

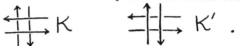

In general, all that is required is that each pair of parallel strands have opposite orientation. Call two knots <u>pass-equivalent</u> if one can be obtained from the other by a combination of ambient isotopy and four-fold switches of this type. For example, the trefoil and the figure eight knot are pass-equivalent:

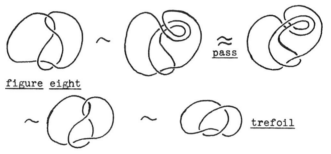

In section 10, when we discuss the Arf invariant, we shall prove the following theorem.

<u>Theorem</u>. Two knots K and K' are pass-equivalent if and only if $a_2(K) \equiv a_2(K')$ (modulo-2). In particular, any knot is pass-equivalent to either the trefoil knot or the unknot.

This theorem provides a direct geometric interpretation for the mod-2 reduction of $a_2(K)$.

9°. Consistency.

Using the switching notation of example 8, let $K' = S_r S_{r-1} \ldots S_1 K$ where K and K' are oriented links. Then repeated application of the exchange identity shows

$$a_{n+1}(K) - a_{n+1}(K') = \Sigma_{i=1}^{r} I_i(K) a_n(E_i S_{i-1} S_{i-2} \ldots S_1 K).$$

Since we can choose a switching sequence so that K' is unknotted or unlinked, this formula suggests that it should be possible to inductively define a_{n+1} in terms of a_n. In [2] Ball and Mehtah take this approach to the Conway polynomial, obtaining an elegant and elementary proof of the consistency of the axioms. The key to their approach is the use of the unknotting procedure illustrated in Figure 24. The interested reader is referred to their paper for further details.

6. Curliness and the Alexander Polynomial

In this section we discuss the structure of a universe viewed as a plane curve immersion. The total turning number of the tangent vector to such an immersion will be given a combinatorial form called here curliness. We then show that Alexander's original algorithm for calculating the Alexander polynomial contains a hidden curliness calculation. It is this calculation that makes the Alexander polynomial dependent, up to factors of the form $\pm t^k$, upon the particular choice of knot projection for a given knot. Thus we show how to normalize the classical polynomial to obtain the state polynomial model for the Conway Polynomial. In the process, we prove the translation formula $\Delta_K(t) \doteq \nabla_K(\sqrt{t} - 1/\sqrt{t})$ relating the Alexander polynomial, $\Delta_K(t)$, and the Conway polynomial. (Here $\doteq$ denotes equality up to factors of the form $\pm t^k$ where k is an integer.)

Immersions

If $\alpha : S^1 \longrightarrow R^2$ is a differentiable mapping of the circle into the plane with non-vanishing differential $d\alpha$, then α is said to be an *immersion*. An immersion α is said to be *normal* if the pre-image $\alpha^{-1}(p)$ of any point $p \in R^2$ is either empty, a single point, or two points $\{p_1, p_2\} = \alpha^{-1}(p)$ with $d\alpha(p_1)$ and $d\alpha(p_2)$ linearly independent. Thus a normal immersion is a locally one-to-one mapping with normally crossing singularities in the image. The same remarks and definitions apply when the single circle is replaced with a collection of disjoint circles Λ. Consequently, any universe U may be represented as the image $\alpha(\Lambda)$ of a normal immersion $\alpha : \Lambda \longrightarrow R^2$, where Λ is a disjoint collection of circles.

Two immersions $\alpha, \beta : \Lambda \longrightarrow R^2$ are said to be regularly homotopic if there is a family of immersions $f_t : \Lambda \longrightarrow R^2$, $0 \leq t \leq 1$ so that $f_0 = \alpha$, $f_1 = \beta$ and f_t and df_t vary continuously with t. Just as ambient isotopy for knots and links is discretized by the three elementary moves (Reidemeister moves) of sections 4 and 5, we obtain a discrete version of regular homotopy via the moves of type A and B illustrated below:

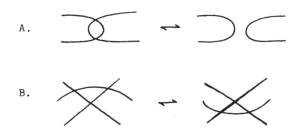

A.

B.

Definition 6.1. Two universes U and U' are regularly homotopic (U ≈ U') if U' can be obtained from U by a sequence of moves of type A and B.

Given an immersion $\alpha : S^1 \longrightarrow R^2$ of an oriented circle into the plane, Whitney [38] defined a degree $D(\alpha) \in Z$. The degree measures the total number of times (counted with sign) that the image unit tangent vector turns through 2π as the curve is traversed once. By convention, we take a counterclockwise circle about the origin to have degree one. Whitney and Graustein proved the fundamental result: Two curves $\alpha, \beta : S^1 \longrightarrow R^2$ are regularly homotopic if and only if they have the same degree.

Using calculus, the degree is defined as follows: Since α is an immersion, $\|\alpha'\| \neq 0$ where $d\alpha = \alpha' d\theta$ (θ is the coordinate on S^1). Let $\beta = \alpha'/\|\alpha'\|$ be the unit tangent vector to α. Then

$$D(\alpha) = \frac{1}{2\pi} \int_0^{2\pi} \beta d\theta .$$

We now give a combinatorial interpretation of the Whitney degree, and relate this interpretation to the states of a universe.

<u>Definition</u> 6.2. When C is an oriented Jordan curve in the plane, define the <u>curliness</u> $\mathscr{C}(C) = \pm 1$ according as the curve is oriented clockwise (+1) or counterclockwise (-1). If U is any universe, let $\hat{U}$ be the collection of Seifert circles for U (see Lemma 3.4 and Figure 13). Thus $\hat{U}$ is obtained from U by splitting all vertices in oriented fashion (compare Figure 30). Define the curliness of U by the formula:

$$\mathscr{C}(U) = \mathscr{C}(\hat{U}) = \Sigma_{C \in \hat{U}} \, \mathscr{C}(C) .$$

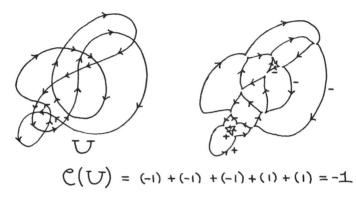

$$\mathcal{C}(U) = (-1) + (-1) + (-1) + (1) + (1) = -1$$

Figure 30

Lemma 6.3. Let $\alpha : S^1 \longrightarrow R^2$ be a normal immersion. Let $U = \alpha(S^1)$ be the corresponding universe. Then the Whitney degree of α coincides with the curliness of U; that is, $D(\alpha) = \mathcal{C}(U)$.

The (easy) proof of the lemma will be omitted. Following our combinatorial theme, it is of interest to give a proof that $\mathcal{C}(U)$ is invariant under regular homotopy. As we see from the next lemma, this analysis will show how $\mathcal{C}(U)$ is related to the topology of Jordan curves in the plane. As usual, a configuration such as ⇌ denotes a universe containing this local pattern, and a formula containing more than one such pattern refers to a single universe that has been changed locally to conform to the indicated patterns.

Lemma 6.4. $\mathcal{C}(\rightleftarrows) - \mathcal{C}( )( ) = 1$

$\mathcal{C}(\leftrightharpoons) - \mathcal{C}( )( ) = -1.$

Proof. It suffices to check these formulas for universes of Jordan curves. The lemma then follows from the chart below and the chart we have not drawn, obtained from the visible chart by reversing all the arrows.

⇄	↓↑	$e(\rightleftarrows)$	$e(↓↑)$
⊜	⊠	2	1
⊘	⊙	0	-1
⤳	Q⊙	-1	-2

Proposition 6.5. If U is regularly homotopic to U', then $\mathcal{E}(U) = \mathcal{E}(U')$.

Proof.

$$\mathcal{E}(\rightleftarrows) = \mathcal{E}(\rightleftarrows) = \mathcal{E}(\supset\subset)$$

$$\mathcal{E}(\rightleftarrows) = \mathcal{E}(\text{⋈})$$

$$= \mathcal{E}(\rightleftarrows) \quad -1$$

$$= \mathcal{E}(↓↑) \quad +1-1 \quad\quad (6.4)$$

$$\therefore \mathcal{E}(\rightleftarrows) = \mathcal{E}(\supset\subset).$$

Thus $\mathcal{E}$ is invariant under moves of type A.

$$\mathcal{E}(\text{⋈}) = \mathcal{E}(\text{⋈})$$
$$= \mathcal{E}(\text{⋈})$$
$$= \mathcal{E}(\text{⋈}).$$

The final case for the type B move is illustrated on the next page.

$$\mathscr{C}\left(\begin{array}{c}\includegraphics\end{array}\right) = \mathscr{C}\left(\begin{array}{c}\includegraphics\end{array}\right)$$

$$= \mathscr{C}\left(\begin{array}{c}\includegraphics\end{array}\right) \quad -1$$

$$= \mathscr{C}\left(\begin{array}{c}\includegraphics\end{array}\right) \qquad (6.4)$$

$$= \mathscr{C}\left(\begin{array}{c}\includegraphics\end{array}\right) +1 \quad (6.4)$$

$$= \mathscr{C}\left(\begin{array}{c}\includegraphics\end{array}\right)$$

$$= \mathscr{C}\left(\begin{array}{c}\includegraphics\end{array}\right) = \mathscr{C}\left(\begin{array}{c}\includegraphics\end{array}\right)$$

This completes the proof of Proposition 6.5.

Any universe is regularly homotopic to a disjoint union of the standard forms indicated in Figure 31. For the reader interested in constructing a combinatorial proof of this fact, the "Whitney trick", as illustrated in Figure 32, will be useful.

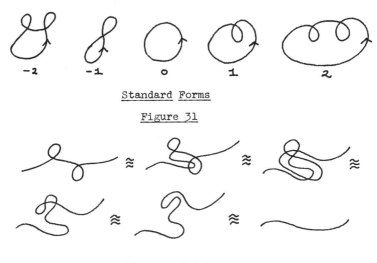

Standard Forms

Figure 31

Whitney Trick

Figure 32

In order to relate the curliness of a universe to its states, it is convenient to use the string form. Upon decomposing a string (see section 2) into Seifert circles, there will be a single string without self-crossings, and a collection of Jordan curves. Assign curliness zero to the string without self-crossings, and the usual plus or minus one to each Jordan curve. Thus

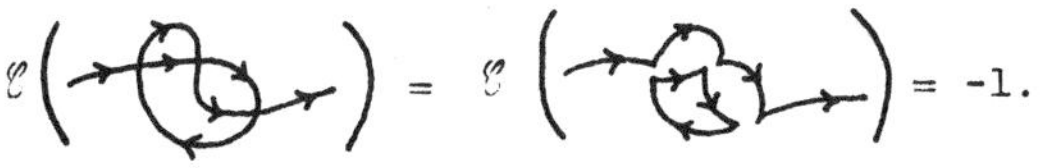

$$C\left(\rightarrow\!\!\bigoplus\!\!\rightarrow\right) = C\left(\rightarrow\!\!\bigoplus\!\!\rightarrow\right) = -1.$$

We now show how to extract the curliness from any state of a universe.

Definition 6.6. A vertical **crossing** label is a label of the form ![vertical crossing with A up-right, D down-left] where $AD = 1$. For a universe U, $V(U)$ will denote the <u>vertical labelling</u> of U obtained by placing a vertical label at each crossing.

Theorem 6.7. Let U be a string universe and $V(U)$ the vertical labelling of U; let S be any state of U. Then

$$\langle V(U)|S\rangle = A^{\mathscr{C}(U)}.$$

Discussion. We first observe that there exist states S of U for which the formula of 6.7 is true. For example:

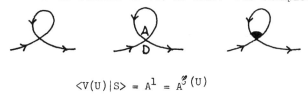

$$\langle V(U)|S\rangle = A^1 = A^{\mathscr{C}(U)}$$

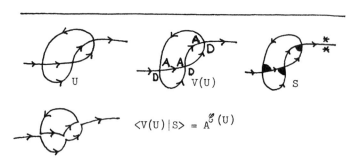

$$\langle V(U)|S\rangle = A^{\mathscr{C}(U)}$$

In order to construct a state for which the formula holds, first form $\hat{U}$, the set of Seifert circles for U. Then create a trail by the following procedure: Locate a Seifert circle with empty interior. Reassemble one of its sites so that the resulting configuration has one fewer circle. Repeat this process, never using a given site more than once. The result is a trail whose corresponding state has the required property. (The proof of this fact will be omitted.) For example:

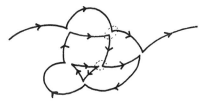

Seifert circles with indicated re-assemblies.

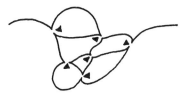

Trail with state markers.

The State S.

Proof of 6.7. By using the relation $AD = 1$ it is easy to see that when S and S' are related by a sequence of state transpositions, then $\langle V(U)|S\rangle = \langle V(U)|S'\rangle$. Thus the theorem follows from the Clock Theorem and the existence of a state S that satisfies the formula.

The Alexander Polynomial

The classical Alexander polynomial (see the appendix) is well-defined up to sign and multiplication by powers of t. In our terms, Alexander uses the following code:

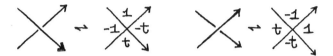

Thus the Alexander polynomial $\Delta_K(t)$, of a knot or link K is defined by the formula $\Delta_K(t) \doteq \mathrm{Det}(A_K(t))$ where $A_K(t)$ is an Alexander matrix (see section 3) for the labelling corresponding to the Alexander code. The symbol, $\doteq$, denotes equality up to factors of the form $\pm t^n$, n an integer.

The proof of the next lemma will be omitted.

Lemma 6.8. The Alexander polynomial in t^2 is given by the formula $\Delta_K(t^2) \doteq \mathrm{Det}(B_K(t))$ where $B_K(t)$ is the Alexander matrix for the code

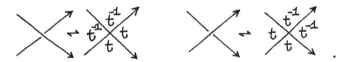

Theorem 6.9. $\Delta_K(t) \doteq \nabla_K(\sqrt{t} - 1/\sqrt{t})$.

Proof. It follows from 6.7 and 6.8 that $\Delta_K(t^2) \doteq t^{-\mathscr{E}(U)} \nabla_K(t)$ where $z = t - t^{-1}$ (take $W = t$, $B = t^{-1}$, $z = W - B$ and the Conway polynomial as defined in Theorem 4.3). This completes the proof.

7. The Coat of Many Colors

We now generalize to link polynomials in many variables. A different variable is assigned to each component of the link. For illustrations we shall use three different types of arrow-head to denote three distinct strands. The new codes are indicated in Figure 33. Note that it is the under-crossing line that determines the code variables.

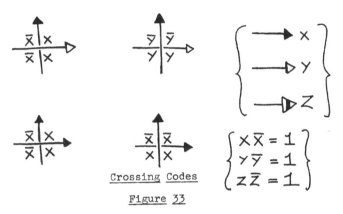

Crossing Codes

Figure 33

Definition 7.1. Let $X_1, X_2, \ldots, X_n$ be the code variables for a link L. Assume that $X_k \bar{X}_k = 1$ for each index k. (Thus these codes are the modified Alexander codes of Lemma 6.8.) Let $\left[X_1^{a_1} X_2^{a_2} \ldots X_n^{a_n} \right]$ denote the difference

$$\left[X_1^{a_1} \ldots X_n^{a_n} \right] = X_1^{a_1} \ldots X_n^{a_n} + (-1)^n \bar{X}_1^{a_1} \ldots \bar{X}_n^{a_n}.$$

As in the single variable case, we have inner products $\langle L | S \rangle$ and state polynomials $\langle L | \mathcal{S} \rangle$ for a given link L whose strands have been labelled with these code variables. However, it is now necessary to give a normalization factor since the unadorned state polynomial is no longer independent of the choice of fixed stars. In order to accomplish this end, we shall use a multiple index for the regions of the universe U. Each region of U is assigned a vector index $p = (p_1, p_2, \ldots, p_n)$. The kth index, p_k, increases or decreases upon crossing the kth strand as indicated in Figure 34. The unbounded region is assigned index $(0, 0, \ldots, 0)$.

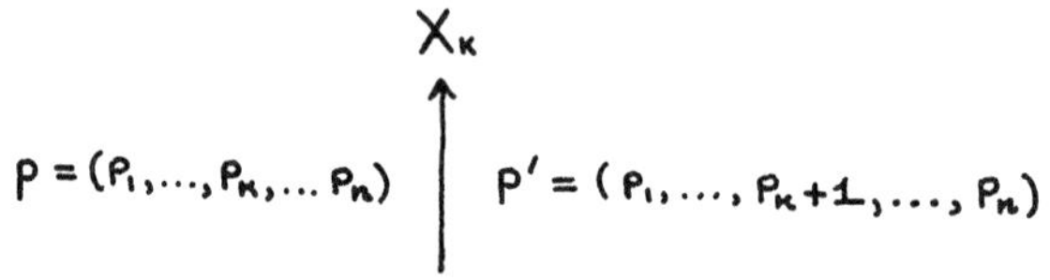

$$X_k \uparrow$$

$$p = (p_1, \ldots, p_k, \ldots p_n) \quad \quad p' = (p_1, \ldots, p_k+1, \ldots, p_n)$$

<u>Multiple Indexing</u>

<u>Figure 34</u>

<u>Proposition 7.2</u>. Let U be a multiply indexed universe corresponding to a link L. Let $\mathcal{S}$ be the collection of states for U whose fixed stars are in a pair of regions with indices $p = (p_1, p_2, \ldots, p_n)$ and $p' = (p_1, \ldots, p_k+1, \ldots, p_n)$ differing only in the kth place. Let $|\mathcal{S}| = (\overline{X}_1^{2p_1} \overline{X}_2^{2p_2} \ldots \overline{X}_n^{2p_n}) \overline{X}_k (X_k - \overline{X}_k)$. Then $\langle L|\mathcal{S}\rangle / |\mathcal{S}|$ is independent of the choice of fixed stars.

The proof of this proposition will be omitted. It follows the same format as the proof of Theorem 3.3.

In order to obtain topological invariance, we must take curliness (see section 6) into account. Given L as above, let $L = L_1 \cup L_2 \cup L_3 \cup \ldots \cup L_n$, the separately labelled sub-links. (Note that we have the option of labelling distinct strands with the same variable.) Then

$$\square_L = \frac{X_1^{\mathscr{C}(L_1)} X_2^{\mathscr{C}(L_2)} \cdots X_n^{\mathscr{C}(L_n)}}{|\delta|} \langle L|\delta\rangle$$

is a topological invariant of L. We refer to $\square_L$ as the potential function of the link L. ([5], [17]).

Example 7.3.

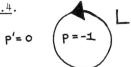

$\langle L|\delta\rangle = 1$ (by convention)
$|\delta| = \overline{X}(X-\overline{X})$, $X^{\mathscr{C}(L)} = X^{-1} = \overline{X}$
$\therefore \square_L = 1/(X-\overline{X})$.

Example 7.4.

$p' = 0 \quad p = -1$

$|\delta| = (\overline{X}^2 {}^{(-1)})\overline{X}(X-\overline{X})$
$\square_L = (X^{\mathscr{C}(L)}/|\delta|)\langle L|\delta\rangle$
$= X/X^2 X^{-1}(X-\overline{X})$
$\therefore \square_L = 1/(X-\overline{X})$.

Example 7.5.

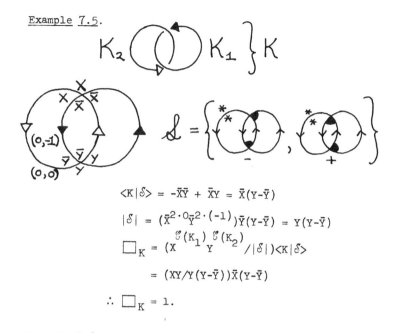

$$\langle K | \mathcal{S} \rangle = -\bar{X}\bar{Y} + \bar{X}Y = \bar{X}(Y-\bar{Y})$$

$$|\mathcal{S}| = (\bar{X}^2 \cdot 0_{\bar{Y}}^2 \cdot (-1)) \bar{Y}(Y-\bar{Y}) = Y(Y-\bar{Y})$$

$$\square_K = (X^{\mathscr{C}(K_1)} Y^{\mathscr{C}(K_2)} / |\mathcal{S}|) \langle K | \mathcal{S} \rangle$$

$$= (XY/Y(Y-\bar{Y}))\bar{X}(Y-\bar{Y})$$

$$\therefore \square_K = 1.$$

Example 7.6. Thus we have

$$\square(\circlearrowright) = 1/(X-\bar{X})$$

$$\square(\circlearrowleft) = 1/(Y-\bar{Y})$$

$$\square(\mathcal{G}) = 1$$

and by a similar calculation,

$$\square(\mathcal{G}) = -1.$$

In general, if L is a single variable link (with variable X assigned to all strands), then

$$(X-\bar{X})\square_L = \triangledown_L$$

where the Conway polynomial is regarded as a function of $z = X - \bar{X}$. In ([5]) Conway uses the notation for the potential function that we have reserved for the Conway polynomial. With this caveat, our normalizations coincide with his.

The potential function still satisfies the basic identity

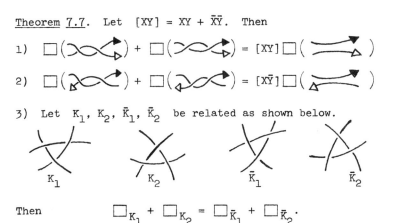

$$\square_K - \square_{\bar{K}} = [X]\square_L$$

for links related at crossings involving identically labelled strands.

In addition to the basic identity there are also identities for crossings of differently labelled strands:

<u>Theorem 7.7</u>. Let $[XY] = XY + \bar{X}\bar{Y}$. Then

1) $\square(\asymp) + \square(\asymp) = [XY]\square(\longrightarrow)$

2) $\square(\asymp) + \square(\asymp) = [X\bar{Y}]\square(\longrightarrow)$

3) Let $K_1, K_2, \bar{K}_1, \bar{K}_2$ be related as shown below.

Then $\quad \square_{K_1} + \square_{K_2} = \square_{\bar{K}_1} + \square_{\bar{K}_2}.$

These weaves can take any consistent orientations, and may be single, dual, or tri-colored (that is, there may be one, two, or three variables involved in the weave).

Proof. Since the potential function is a normalized state summation, these identities can be demonstrated by the same method as used in Theorem 4.1. In Figure 35 we have shown all of the local state configurations with their contributions to the terms of state summations for the first identity (1). Four configurations cancel in pairs, leaving the pattern of this identity. The second identity follows in the same fashion. The verification of the third identity is a long calculation involving the triangle states (Figure 23) and is omitted.

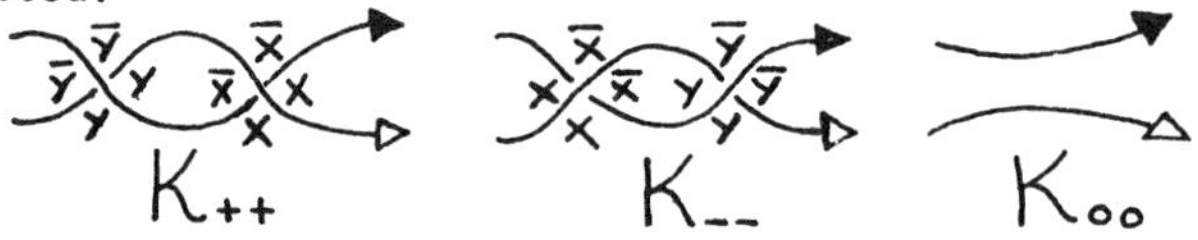

++ Term	-- Term	Local State
$\bar{y}\bar{x}$	xy	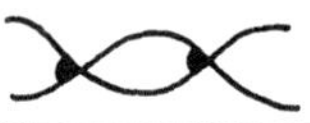
$\bar{y}\bar{x}$	$\bar{x}y$	
$y\bar{x}$	$\bar{x}\bar{y}$	
$y\bar{x}$	xy	
yx	$\bar{x}y$	
yx	$\bar{x}\bar{y}$	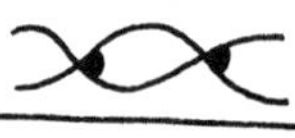

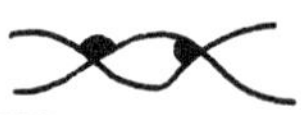

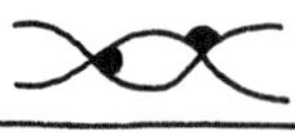

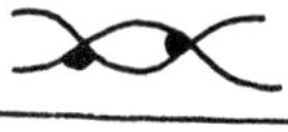

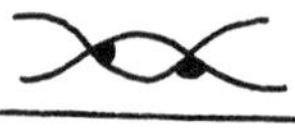

Figure 35

Remark. These identities do not yet provide an axiomatization for the potential function. It appears to be necessary to include the values of the potential function on a large class of links (calculated via state enumeration or by other methods) in order to determine the potential for all links. This must be contrasted with the situation for the Conway polynomial. The Conway polynomial is determined for all knots and links by its axioms once we know the value of the unknot. How large a collection of links must we specify in order to determine the potential recursively from the identities and the fact that ☐ vanishes for split links?? In Conway's terms, <u>what are the generators of the polychrome skein</u>?

Proposition 7.8. Let L' be obtained by linking a strand of the link L with an unknotted circle of linking number $\lambda = \pm 1$ as shown below. Then $☐_{L'} = \lambda[Y]☐_L = \pm[Y]☐_L$ where Y is the variable labelling the strand of L that is linked by the new component.

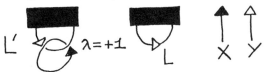

Proof. Apply the definitions to the states and labels indicated below. The details of the calculation are omitted.

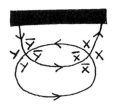

Example 7.9. Let L_n be obtained from L by adding a new strand with linking number n, as shown below.

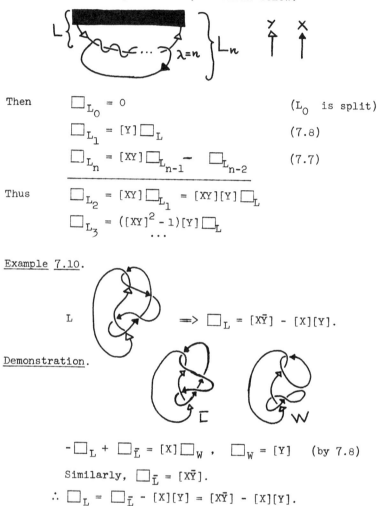

Then $\square_{L_0} = 0$ (L_0 is split)

$\square_{L_1} = [Y]\square_L$ (7.8)

$\square_{L_n} = [XY]\square_{L_{n-1}} - \square_{L_{n-2}}$ (7.7)

Thus $\square_{L_2} = [XY]\square_{L_1} = [XY][Y]\square_L$

$\square_{L_3} = ([XY]^2 - 1)[Y]\square_L$

...

Example 7.10.

$\Longrightarrow \square_L = [X\bar{Y}] - [X][Y]$.

Demonstration.

$-\square_L + \square_{\bar{L}} = [X]\square_W$, $\square_W = [Y]$ (by 7.8)

Similarly, $\square_{\bar{L}} = [X\bar{Y}]$.

$\therefore \square_L = \square_{\bar{L}} - [X][Y] = [X\bar{Y}] - [X][Y]$.

Example 7.11.

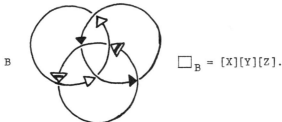

$\square_B = [X][Y][Z]$.

B is known as the Ballantine rings. We omit the calculation of its potential function. (There are sixteen states in the Ballantine Universe.)

Example 7.12.

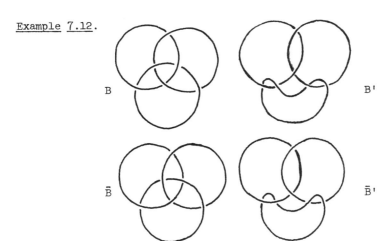

Let this family inherit color and orientation from the Ballantine rings in Example 7.11.

$\square_{\bar{B}} = -[XYZ]$ (calculation omitted).

$\square_{\bar{B}'} = 0$ (split).

$\square_B + \square_{B'} = \square_{\bar{B}} + \square_{\bar{B}'}$ (Theorem 7.7).

$\therefore \square_{B'} = -([XYZ] + [X][Y][Z])$.

8. Spanning Surfaces

Let S^3 denote the (oriented) three-dimensional sphere. An oriented surface F embedded in S^3 is said to be a <u>spanning surface</u> for a knot or link K if K is identical with the boundary of F, and the orientation on F induces the given orientation on K.

In [36] Seifert gave a specific method for constructing a spanning surface associated with a given knot diagram. His surface will be designated the <u>Seifert surface</u> of the diagram. In this section we shall describe Seifert's algorithm, relate the degree of the Conway polynomial to the genus of the Seifert surface, and discuss the generalization of these relationships to arbitrary spanning surfaces via the Seifert pairing.

Seifert's Algorithm

Form the Seifert circles (see Figure 13) on the knot diagram by drawing two arcs at each crossing as indicated below.

The Seifert circles are then obtained by traversing the diagram along its edges and moving along the arcs when arriving at a crossing. Thus each circle is embedded (in the plane) in the complement of the crossings. The Seifert surface is then constructed by attaching discs to the Seifert circles so that these discs are embedded disjointly in the three space including the plane of the diagram and extending above it. The surface is completed by filling in a twisted band at each crossing, as shown below.

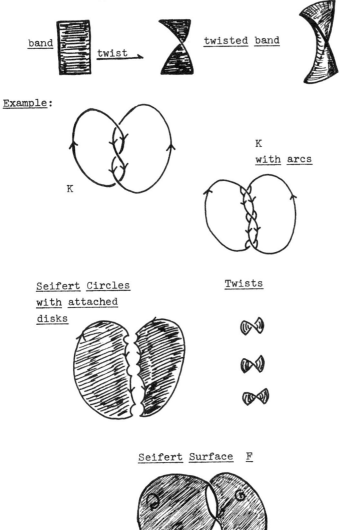

Example:

K

K with arcs

Seifert Circles with attached disks

Twists

Seifert Surface F

Note that the Seifert circles constructed in this process are the same as those produced by splitting sites to preserve orientation. Thus, in the example above, the circles drawn by site-splitting appear as

U   Seifert Circles

Each twisted band has the same homotopy type as its projected image in the plane, a filled-in crossing:

Consequently, the Seifert surface has the homotopy type of the 2-cell complex obtained, from the universe U underlying the knot diagram, by adding disjoint 2-cells to the Seifert circles. (Regard these cells as automatically filling in the crossings.)

2-cell complex

Lemma 8.1. Let K be a knot or link (diagram) with underlying universe U. Let F be the Seifert surface for the diagram K, and let $\chi(F)$ denote the Euler characteristic of F, and $\rho(F)$ denote the rank of the first homology group $H_1(F)$. Then

$$\chi(F) = S(U) - R(U) + 2,$$
$$\rho(F) = R(U) - S(U) - 1,$$

where $R(U)$ = the number of regions in U, and $S(U)$ = the number of Seifert circles in U. Note that the Euler characteristic and rank are functions of the universe alone.

Proof. Let $V = V(U)$ denote the number of vertices of U, and $E = E(U)$ denote the number of edges of U, while $R = R(U)$ and $S = S(U)$. Then $V - E + R = 2$ since the plane has Euler characteristic 2, while the Euler characteristic of the Seifert surface is given by $\chi(F) = V - E + S$ since the surface has the homotopy type of the two-cell complex constructed on the universe. (Recall that the Euler characteristic of a two-complex is equal to $N_0 - N_1 + N_2$ where N_i is the number of i-cells in the complex. Recall also that for a connected surface with boundary, the Euler characteristic equals $2 - \rho$ where ρ is the rank of the first homology group.) The lemma follows at once from these formulas.

Remark. A surface with boundary is said to have genus g if it becomes a sphere with g handles upon adding disks to all the boundary components. This gives rise to the relation

$\rho = 2g + \mu - 1$ for a connected surface with μ boundary components. Hence we have the

Corollary 8.2. Let K be a link diagram with μ components, connected universe U, R regions, and S Seifert circles. Then the Seifert surface F for K has genus g given by the formula $g = \frac{1}{2}(R - S - \mu)$.

Remark. It is of interest to note that a surface of relatively low genus can arise from a diagram with many regions if these regions are balanced off by a goodly collection of Seifert circles. The simplest case is an unknot of the form

R = 10
S = 9 $g = \frac{1}{2}(10 - 9 - 1) = 0.$
μ = 1

Remark. Lest we give the impression that the Seifert surface is easy to draw in perspective, please note that nested Seifert circles give rise to nested disks. It is a well-known bit of folklore that any knot has a diagram in its ambient isotopy class without nested Seifert circles. I leave it to the reader to make this concept of nesting precise, but point out that of the two equivalent diagrams shown below for the figure eight knot, only the second has a Seifert surface that can be drawn without overlapping disks (allowing one disk to be indicated by an unbounded planar region).

The Degree of the Conway Polynomial

Call a site of the form ⤫ an <u>active site</u> and a site of the form ⤦⤧ a <u>passive site</u>. We adopt this terminology because, in the state corresponding to a trail, the black and white holes of the state are in one-to-one correspondence with the active sites of the trail.

In order to determine an upper bound for the degree of the Conway polynomial, we shall use the state polynomial model for $\nabla_K(z)$. Thus $\nabla_K(z) = \langle K | \mathcal{S} \rangle$ with the caveat that $BW = 1$ and $z = B - W$. (See section 4.) Thus the degree of $\nabla_K(z)$ is equal to the maximal power of B in the state polynomial $\langle K | \mathcal{S} \rangle$ reduced by the relation $BW = 1$. By the definition of the state polynomial, this maximal power can be no more than the largest number of active sites possible for a trail on the universe underlying K.

Lemma 8.3. Let U be a universe with R regions and S Seifert circles. Then any trail on U has no more than $R - S - 1$ active sites.

Proof. Split every vertex of U to form an active site. This creates the collection of Seifert circles $\mathcal{C}$. $\mathcal{C}$ can be converted into a trail in no less than $S - 1$ re-assemblies. By using this many reassemblies, the trail has the maximal number, A, of active sites. Thus $A = V - (S-1)$ where V is the number of vertices of U. Since $V = R - 2$, $A = (R-2) - (S-1)$. Q.E.D.

Proposition 8.4. Let K be a knot or link (diagram) with connected underlying universe U. Let F be the Seifert surface for K, and ρ = rank $H_1(F)$. Then

$$\deg \nabla_K(z) \leq \rho.$$

Proof. By Lemma 8.3 and the remarks preceding it, $\deg \nabla_K(z) \leq A = R - S - 1 = \rho$ by Lemma 8.1. This completes the proof.

Remark. When $\deg \nabla_K = \rho$, then we know that the given Seifert surface has minimal genus among all Seifert surfaces spanning other diagrams for the knot. For example, let K be the knot shown below. Then $\rho = 4$, and $\nabla_K = 1 - z^2 - z^4$. Thus the Seifert surface for K has minimal genus.

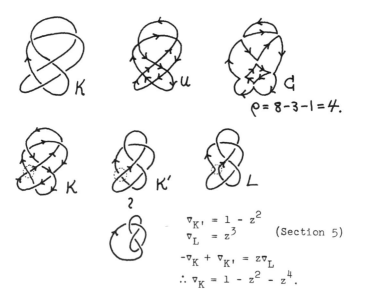

Drawing the Seifert Surface

Each Seifert circle divides the plane into two regions. Call a Seifert circle of <u>type I</u> if one of these regions is void of other Seifert circles; otherwise call the Seifert circle of <u>type II</u>.

In order to understand the structure of Seifert surfaces involving type II circles it is convenient to have a picture of the surface near its boundary. The difficulty is illustrated by trying to see this for the figure eight knot:

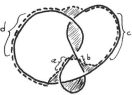

The dotted circle will have a disk attached to form the surface. This disk is out of the plane, attached perpendicularly along the dotted curve. We do not yet see a collar of the boundary of the surface along segments a, b, c, d. To rectify this consider the following diagram (Figure 36).

Figure 36

In Figure 36 the dotted circle has been replaced so that it forms a new component in the diagram that over-crosses any old component that it meets. If the old diagram is K, let $\hat{K}$ denote this new diagram. As Figure 36 demonstrates, $\hat{K}$ has only type I Seifert circles. The Seifert surface F_K is obtained from $F_{\hat{K}}$ by adding disks to the dotted circles (the new components) in $F_{\hat{K}}$. (Note that each new over-crossing component is endowed with an orientation opposite to that of its Seifert circle of origination.) Thus we have an embedding $F_{\hat{K}} \subset F_K$. The surface $F_{\hat{K}}$ depicts the planar portion of F_K in a graphic and useful manner.

Example. Here is a sequence of drawings leading to a depiction of the (minimal) Seifert surface for the knot discussed just after Proposition 8.4. In the final drawing the Seifert surface is shown with one of the disks occupying the unbounded portion of the plane. It is obtained from the previous picture by swinging the upper (twisted) band underneath the rest of the diagram.

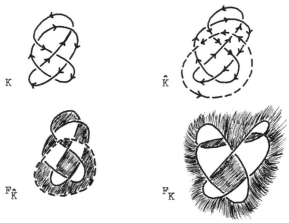

Arbitrary Spanning Surfaces

Proposition 8.4 can be generalized to arbitrary spanning surfaces. That is, for any connected, oriented surface $F \subset S^3$ spanning a link K, deg $\nabla_K(z) \leq \rho(F)$ where $\rho(F)$ is the rank of $H_1(F)$. In order to see this generalization, we need another model of the Conway polynomial. This model is discussed in [21]. I will outline the result here.

Given a spanning surface $F \subset S^3$, there is a pairing (<u>The Seifert pairing</u>) $\theta : H_1(F) \times H_1(F) \longrightarrow Z$ defined by the formula $\theta(a,b) = Lk(a^+,b)$ where $Lk(,)$ denotes linking number in S^3, and a^+ is the result of translating the cycle a into $S^3 - F$ along the positive normal to F. The Seifert pairing is an invariant of the ambient isotopy class of the embedding of the surface in the three-sphere. It can be used to create invariants of the embedding of the boundary of this surface.

Let $\Omega_K(X) = D(X\theta - X^{-1}\theta^T)$ where D denotes determinant, θ denotes a matrix of the Seifert pairing with respect to a basis for $H_1(F)$, and T denotes matrix transpose. Then, letting $z = X - X^{-1}$, one has that $\nabla_K(z) = \Omega_K(x)$. This means that the Conway polynomial can be computed via the Seifert pairing from any spanning surface for the link. Since the Seifert matrix has size $\rho(F) \times \rho(F)$, the rank inequality follows immediately.

<u>Example</u>. Let F be the surface shown below. It is easy to see that K, the boundary of F, is isotopic to the trefoil knot. $H_1(F)$ is generated by a and b so that the Seifert

matrix is $\begin{bmatrix} -1 & 0 \\ 1 & -1 \end{bmatrix}$ with respect to this basis. That is, $\theta(a,a) = \theta(b,b) = -1$ and $\theta(a,b) = 0$, $\theta(b,a) = 1$.

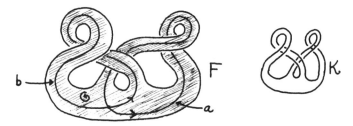

Thus
$$\Omega_K(X) = D\left(X\begin{bmatrix} -1 & 0 \\ 1 & -1 \end{bmatrix} - X^{-1}\begin{bmatrix} -1 & 1 \\ 0 & -1 \end{bmatrix}\right)$$

$$= D\begin{bmatrix} -X+X^{-1} & -X^{-1} \\ X & -X+X^{-1} \end{bmatrix}$$

$$= (-X+X^{-1})^2 + 1$$

$$= (X-X^{-1})^2 + 1$$

$$\Omega_K(X) = z^2 + 1 = \nabla_K(z).$$

Remark. It is curious to note that with $z = X - X^{-1}$ we have $X = z + 1/X$, hence $X = z + \cfrac{1}{z + \cfrac{1}{z + \cfrac{1}{z + \ldots}}}$

and
$$\nabla_K(z) = \Omega_K\left(z + \cfrac{1}{z + \cfrac{1}{z + \cfrac{1}{z + \ldots}}}\right).$$

In particular $\nabla_K(1) = \Omega_K\left(\dfrac{1+\sqrt{5}}{2}\right)$.

9. The Genus of Alternative Links.

In this section we define a class of alternative links and prove that the Seifert surface of an alternative link diagram has minimal genus among connected spanning surfaces for the link in three-dimensional space. This theorem generalizes a corresponding result by Crowell and Murasugi for alternating links [32]. [8,28,29] It is also related to the work of Murasugi and Mayland on pseudo-alternating links [8]. [32] Alternative links are pseudo-alternating, and we conjecture that these two classes of links are identical to one another.

An alternating link has a weaving pattern that is instantly recognizable from a suitable diagram. In such a diagram the over and under crossings alternate as one traverses the components of the link. For example, the trefoil knot, the Ballantine rings, and the example just after Proposition 8.4 are all alternating.

An alternative diagram also has a recognizable weaving pattern, but this pattern does not become apparent until the diagram is translated into the form of the Seifert circles with an appropriate coding. For this reason, I first consider codes and designations for crossings in a knot diagram. Figure 37 lists the different notations that we shall use. Row (a) gives the diagrammatic or pictorial representation. Row (B) shows the well-known [1] double-dot convention. Two dots are placed to the left of the under-crossing line, one on each side of the over-crossing line. Row (C) is a variant of Row (B) -- The crossing has been split to form an active site (see section 8 for the definition of active and passive sites) and one dot appears just to the left or to the right of the site. I call this the <u>site marking code</u>. Row (D) exhibits the label code that we have used for calculating state polynomials.

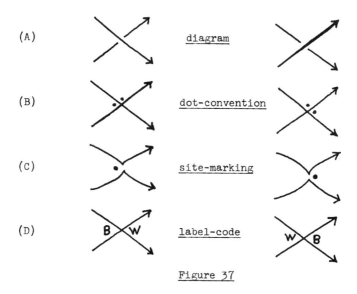

Figure 37

By labelling a diagram with the dot convention it is easy to translate to the site-marking code. Note that the site-marking code appears as the collection of Seifert circles decorated with dots at the sites. For example:

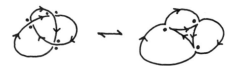

Observe that the Seifert circles divide the plane into connected sets that we call <u>spaces</u> (as opposed to the <u>regions</u> of the knot diagram). Thus the trefoil diagram has five regions, while the corresponding set of Seifert circles has three spaces. In a checkerboard coloring of the knot diagram each space in the diagram of Seifert circles receives a solid color. Spaces that receive the same color will be said to have the same <u>parity</u>.

Example. 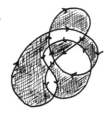  <u>spaces of same parity have same color</u>.

checkerboard

<u>Definition 9.1</u>. Let K be a link with connected under-lying universe, and CK be the diagram of Seifert circles for K decorated according to the site marking convention (Figure 37). K is said to be an <u>alternative link</u> if all marks in any given space of CK have the same type. When speaking of links up to topological equivalence, we shall say that a link is alternative if there is an alternative diagram in its topological equivalence class.

In Figure 38 we give an example of an alternating knot and show that it is alternative. On the other hand, there are knots and links that are not alternating but nevertheless alternative. The same figure illustrates this for the (3,4) torus knot.

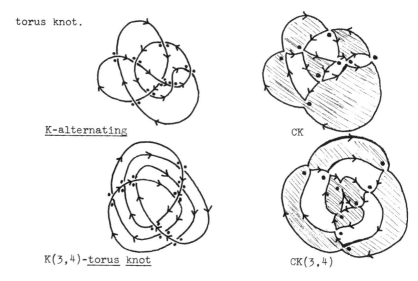

Figure 38

Alternating links are alternative links with the following special property:

Lemma 9.2. A link diagram is alternating if and only if it is alternative and spaces in the diagram of Seifert circles receive the same or different marking type according to whether they have the same or different parity.

The evidence for this lemma can be seen from the example in Figure 38. Figure 39 illustrates the key geometry that is relevant for the proof of the lemma. Further details are omitted.

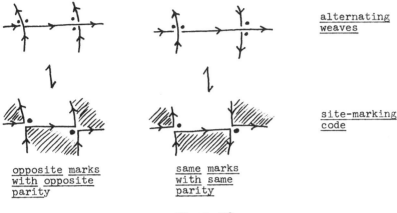

alternating weaves

site-marking code

opposite marks with opposite parity

same marks with same parity

Figure 39

We are now prepared to state the main theorem of this section.

Theorem 9.3. Let L be an alternative link diagram, F_L the Seifert surface for L. Then F_L is a minimal genus spanning surface for L. In fact, deg $\nabla_L = \rho(F_L)$ where $\rho(F_L)$ is the rank of the first homology group of F_L.

This is the promised result about alternative links. In order to prove it, we must locate those states S of the universe underlying L that yield highest B-power inner products $\langle L|S \rangle$. The key observation is due to Ivan Handler [16]: When the site-marking code and the label code are superimposed, the dot always falls on the B. See Figure 40.

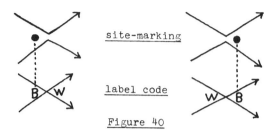

Figure 40

The Moral: To obtain a maximal B-power state S, reassemble sites on CL (the diagram of Seifert circles) so that there are a maximal number of active sites, and in the associated state the state marker falls on the site marking dot at each active site.

Each such state marker will then contribute a B to $\langle L|S \rangle$. The state S will have $\deg\langle L|S \rangle = \rho(F_L)$ exactly when there are $S_L - 1$ reassemblies, and every state marker at an active site (i.e., a site that has not been assembled) falls on a dot.

We now describe an algorithm that creates high power states, and give examples of its application.

Alternative Tree Algorithm (ATA)

1. Let L be an alternative projection. Let CL denote the Seifert circles of L decorated with the site marking dots that code for L. By definition, each space of CL has markings of a single type. Choose stars in a pair of adjacent regions of L.

2. Grow a tree rooted at a star. The tree can branch from one region of L to another if and only if
   a) The second region is unoccupied by tree branches.
   b) There is a site that opens from the first region to the second.
   c) At an active site, the marking dot must be in the second region.

branch   grow

Grow trees from each star until a), b), and c) can no longer be satisfied.

3. There will now be a collection of sites (at boundaries of regions occupied by tree-branches) without any branches passing through them. (Further branching being forbidden by the restrictions of 2.) These sites will be reassembled. Denote the reassembly by placing a circle around the site.

The trees now have new access. Continue growing them according to part 2., until no further growth is possible. Apply 3. again and continue in this manner until each site has a branch passing through it.

4. Reassemble the boxed sites and use the two trees to create a state S. This state will satisfy $\rho(F_L) = \deg \langle L|S \rangle$. There will be $S_L - 1$ reassemblies, and every state marker will fall on a dot.

In this algorithm each stage of growth involves branching choices. Thus many different states can appear as the end product. However, <u>if S and S' are states obtained by ATA from the same link diagram L, then S and S' have the same sign</u>: $\sigma(S) = \sigma(S')$.

<u>Reason</u>. Different branching choices always occur within the same space of CL and will therefore leave the total number of black and white holes in the corresponding states invariant, since all markers in a given space are of the same type.

By the dint of this sign constancy, the polynomial ∇_K has a term $\pm N \cdot B^{\rho(F_L)}$ where N is the number of different states produced by the algorithm ATA. Theorem 9.3 follows once it is shown that N is not equal to zero, and that the algorithm produces all of the maximal power states. This is true, and will be verified after we give a series of examples that illustrate how the algorithm works.

<u>Example 9.1</u>. Let K be the figure-eight knot.

Then K is alternating, and the diagram for CK shows the dots of opposite type in spaces of opposite parity. By choosing stars as shown below, the first stage of ATA yields unique trees and reassemblies:

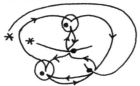

In this case, one more pass through ATA completes the process:

Here is the resulting trail:

If S denotes the state corresponding to this trail, then S has one black hole and one white hole. Hence $\sigma(S) = -1$. The marker at each of the active sites falls on a dot, and hence on a B in the label diagram for K. Hence $\langle K|S \rangle = B^2$, and S is a maximal B-power state for K.

Note that in computing this example the end result is actually obtained (just before the last pass) when all the reassemblies have been indicated. In future computations we shall stop the process at this point.

Example 9.2. Let L be the alternating link shown below.

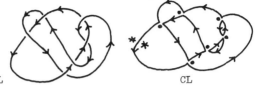

In this case, the first pass through ATA produces two maximal states:

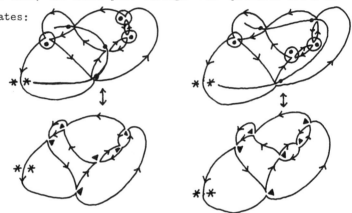

Example 9.3. Let L be the (3,3) torus link as shown below. This diagram is alternative, but not alternating.

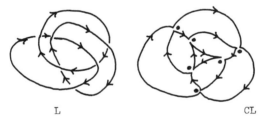

In this case, the algorithm produces a unique state in two passes:

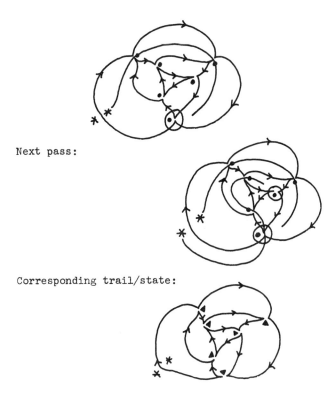

Next pass:

Corresponding trail/state:

Example 9.4. Let K be the alternating knot shown below.

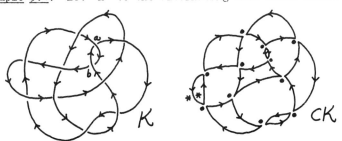

In this case, one can see that ATA constructs 11 maximal states.

If we let K^* be the knot obtained by switching crossings a and b (labelled above), then the resulting projection is not alternative. The algorithm can still be used to enumerate the maximal states of K^*. Of these states, 7 have negative signs and 4 have positive signs. Hence after cancellation there is a coefficient of -3 on the highest power term, and degree $\nabla_{K^*} = \rho(F_{K^*})$. We conjecture that K^* is not alternative in any projection.

This example shows how these methods may be used to show that apparently non-alternative knots may achieve minimal genus on a Seifert surface. The problem of showing that such a knot does not possess an alternative projection is very difficult.

Why ATA produces all the high power states.

We now explain why the algorithm ATA produces all the high power B-states for an alternative link L. Recall that if CL has s Seifert circles, then each maximal trail/state is obtained from CL by reassembling s-1 sites (not all such reassemblies yield maximal states.).

In order to understand the geometry of the reassembly it is helpful to phrase it more generally: The set of Seifert circles is a special case of an arbitrary collection of disjoint circles in the plane. Given two circles, let a reassembly between them be denoted by a straight line segment drawn from one to the other. Thus

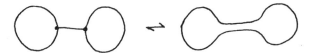

Given s circles in the plane, we can ask how many reassembly lines need be drawn to create a simple closed curve. The answer, of course, is s - 1 just as for the special case of Seifert circles. In the general case we are free to reassemble wherever we please, rather than at specified sites. For example:

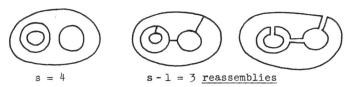

s = 4   s - 1 = 3 <u>reassemblies</u>

Where there are a minimal number of reassemblies (i.e., s - 1 reassemblies for s circles) then <u>each space remains connected after deletion of the reassembly lines</u>. For example, in the diagrams above, one of the spaces is a disk with two holes -- after cutting it along the reassembly lines it becomes a disk:

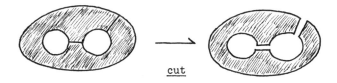

<u>cut</u>

This transformation of spaces to disks must occur if the re-

assembly is to produce a Jordan curve in the end. To obtain
such a curve with the least number of reassemblies it suffices
to choose a set of reassemblies in each space that will cut
that space to a disk.

Let TA denote the algorithm ATA without stipulation
(2c) (directedness of the tree). Then TA chooses a set of
reassemblies in each space by growing a tree in that space --
if we translate to the language of circles and reassembly lines
it becomes clear that this produces a set of reassembly lines
that cut the space to a disk. Thus TA will produce every
state that corresponds to s-1 reassemblies.

The algorithm ATA is designed to select exactly those
states produced by TA where every marker at an active (not
reassembled) site falls on a dot. In order to do this ATA
must grow a tree in each space so that the reassemblies cut
the space to a disk. That this can be done follows from a
graph theoretic lemma:

Definition 9.4. A graph is said to be directed if each edge
has an assigned orientation. A finite, connected, directed
graph is said to be even if:

1) Each vertex touches an even number of edges.
and 2) At a given vertex, half of the edges are outwardly
directed, and half are inwardly directed (that is,
away from, and toward the vertex respectively).

A vertex v in a graph G is said to be the root of an
oriented maximal tree in G if v is contained in a maximal
tree in G and every vertex in G can be reached from v

by an outwardly oriented path in the tree.

Lemma 9.5. Let G be an even graph and v a vertex of G. Then v is the root of an oriented maximal tree $T \subset G$ (orientation induced by G).

Proof. Since an even graph cannot be itself a tree, it must contain cycles. Hence, by the definition, it contains oriented cycles. Let $A \subset G$ be an oriented cycle. Let $G' = G/A$ be the graph obtained by collapsing A to a point. That is, G' is obtained by removing all edges in A and identifying all vertices in A to a single vertex. Note that G' still satisfies the hypotheses of the lemma (It may be a single vertex.). We say that G is obtained by blowing up a vertex of G'. We may assume by induction that the lemma is true for G'. If T' is a maximal oriented tree for G', then we obtain a tree T for G by lifting T' to G, and extending it by traveling around the cycle (See Figure 41). This completes the proof of the lemma.

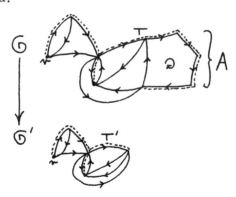

Figure 41

We apply this lemma as follows: Let L be an alternative link diagram, and let CL be the collection of Seifert circles of L marked by the dot-marking convention. Let P be a space of CL. Then P is divided into a number of regions, accessible to each other by passing through the cusps of sites. Assign a graph G(P) to P via:

1) The vertices in G(P) are in one-to-one correspondence with the regions of P.

2) Two vertices are connected by an edge of G(P) whenever the corresponding regions are accessible to one another via a site. The edge is directed toward the region that contains the dot that marks the site.

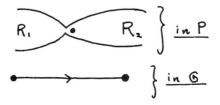

It is easy to see that when all the dots in P are of the same type (as they must be for an alternative diagram) then the graph G(P) is even. This follows from the fact that interacting Seifert circles are oriented coherently. Hence the lemma applies to G(P). This guarantees that the tree-growth demanded by ATA is actually possible.

Thus we have proved that ATA produces a non-empty set of states. This completes the proof of Theorem 9.3.

Examples and Comments

Example 9.5. The knot 8_{19} of the knot tables [33] is standard, hence alternative, but it is not alternating [9]. This is the simplest example of an alternative knot that is neither a torus knot nor an alternating knot.

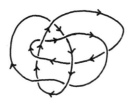

8_{19}

Example 9.6. The Lorenz links studied by Birman and Williams [3] are standard, and hence alternative. These links arise as knotted periodic orbits in dynamical systems on three space. Lorenz links include torus knots and algebraic knots (an algebraic knot is obtained as the link of a plane curve singularity [27]).

Example 9.7. Murasugi has proved that the coefficients of the Alexander polynomial of an alternating knot are all non-zero (i.e., there are no degree gaps) and that they alternate in sign (See [30]). It is a good exercise to prove this theorem using our methods. The alternation is then seen to rest in the change in state signs under transpositions.

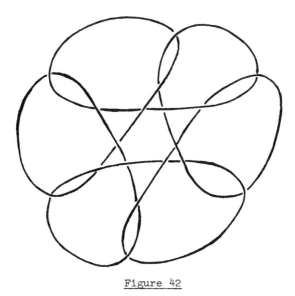

Figure 42

Exercise 9.8. Is the alternative knot shown in Figure 42 alternating in some projection?

10. Ribbon Knots and the Arf Invariant

In this section we examine the notion of pass-equivalence (section 5, remark after Theorem 5.7) more carefully, and show its connection with the detection of ribbon knots and with the Arf invariant. We conclude with a short proof of the theorem of J. Levine, relating the Arf invariant of a knot to the value of its Alexander polynomial at minus one (taken modulo eight).

Definition 10.1. A ribbon knot $K \subset S^3$ is a knot that bounds an immersed disk $\Delta \subset S^3$ having only <u>ribbon</u> <u>singularities</u>. A ribbon singularity is a transverse self-intersection of the immersed disk Δ along an arc s so that if the immersion is represented by a map $\alpha : D^2 \longrightarrow S^3$ ($\alpha(D^2) = \Delta$) then $\alpha^{-1}(s)$ consists of two arcs on D^2, one contained in the interior of D^2, the other touching the boundary of D^2 transversely at its endpoints. Each arc is nonsingularly embedded in D^2.

A typical ribbon singularity is illustrated in Figure 43.

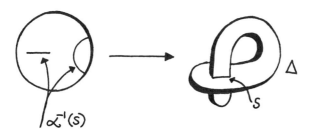

Figure 43

Figure 44 shows that the connected sum of a trefoil and its mirror image is ribbon. I have given two examples of a ribbon disk for this knot. The first should enable the reader to prove that the connected sum of <u>any</u> knot with its mirror image is ribbon.

<u>Remark</u>. A ribbon knot is an example of a <u>slice knot</u>, where this term refers to a knot $K \subset S^3$ that bounds a smoothly embedded disk in the four-dimensional ball, D^4, whose boundary is S^3. This notion was introduced in the paper by Fox and Milnor [13]. It is conjectured that all slice knots are ribbon (see [12]). In accordance with our combinatorial theme, we shall restrict our attention here to ribbon knots. They will be viewed as knots whose projections admit a special weaving pattern described by the ribbon disk.

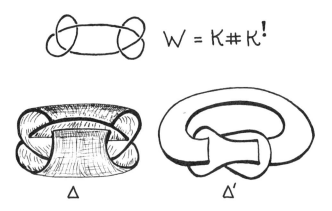

Figure 44

Here is a formal statement of the result about mirror images:

__Lemma 10.2__. Let K be a knot, and let $K^!$ be its mirror image (obtained from a diagram for K by reversing all crossings). Then the connected sum, $K \# K^!$, of K and its mirror image is a ribbon knot.

__Proof__. The ribbon disk is obtained by joining points on the knot and its mirror companion as shown in Figure 44.

__Remark__. This result is due to Fox and Milnor [13]. It forms the basis for the construction of the knot concordance group (See [13], [24], [4]).

The rest of this chapter is devoted to the study of the Arf invariant of a knot. This invariant was discovered by Robertello in [34], and has been examined and reformulated by many authors (See particularly the paper [25] by J. Levine). Our approach to the Arf invariant is geometrical, based on certain constructions in Robertello's original paper. We show that under pass equivalence all knots fall into two distinct classes, and that ribbon knots fall into a class containing the trivial knot. The Arf invariant identifies the pass-class of the knot, and thus can be used to show that certain knots are not ribbon. A more sophisticated (and more algebraic) formulation of the Arf invariant, using the Seifert pairing, shows that it vanishes on any slice knot.

Pass Equivalence

Recall that two knots K_1 and K_2 are said to be <u>pass-equivalent</u> ($K_1 \approx K_2$) if one can be obtained from the other by ambient isotopy combined with a sequence of <u>pass-moves</u>. A pass-move is a local operation on the diagram in one of the following two forms:

1.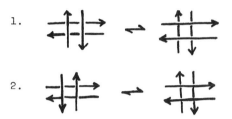

2.

<u>Notation</u>. The symbol $\underset{a}{\approx}$ will be used for ambient isotopy while $\underset{p}{\approx}$ will denote pass-moves.

The next lemma is the key ingredient in showing that ribbon knots are pass-equivalent to the unknot. It is due to Bob Brandt, Thaddeus Olzyk, and Steve Winker (all students in a knot theory seminar held at the University of Illinois at Chicago in 1982).

<u>Lemma 10.2</u>. a)

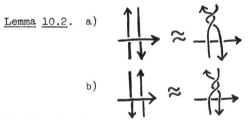

b)

Single strand pass between parallel strands causes $360°$ twist.

Proof of Lemma 10.2.

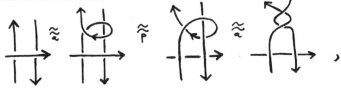

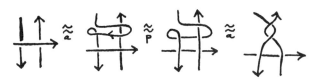

Theorem 10.3. If K is a ribbon knot, then K is pass-equivalent to the unknot: K ≈ 0.

Proof. By a suitable ambient isotopy the ribbon singularities for a ribbon disk spanning K may be exhibited in the local forms:

(The proof of this statement is omitted.)

With this presentation of the ribbon disk, use the Lemma to pass K to a diagram free from ribbon singularities:

Then K ≈ K' where K' bounds a non-singular disk. Hence K' is ambient isotopic to the unknot. Therefore K is pass-equivalent to the unknot.

Q.E.D.

Figure 45 illustrates an ambient isotopy of the stevedore's knot to a ribbon knot, and a passage of this ribbon to the unknot. This is an example of a ribbon knot that is not the connected sum of a knot and its mirror image. (The Conway polynomial of the stevedore is irreducible, while the polynomial of any non-trivial connected sum is a non-trivial product.) This example should underline the extraordinary subtlety in deciding whether a given knot is ribbon.

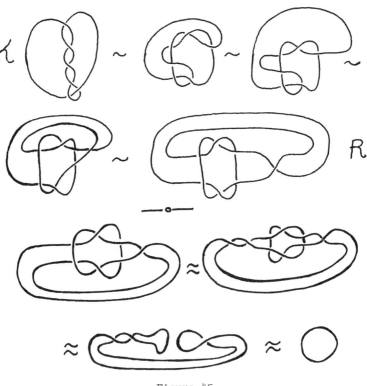

Figure 45

Corollary 10.4. The trefoil is not ribbon.

Proof. According to Theorem 5.7 of these notes, the second degree coefficient, $a_2(K)$, of the Conway polynomial of a knot K is an invariant of pass-equivalence when taken modulo-two. Since a_2(unknot) equals zero, while a_2(trefoil) equals one, we conclude that the trefoil has no passage to the unknot. If the trefoil were ribbon, it would be pass-equivalent to the unknot. This completes the proof.

Definition 10.5. The Arf invariant, $A(K)$, of a knot K, is the value of $a_2(K)$ (see above) taken modulo-two.

This is not the historical definition of the Arf invariant. It is the most convenient definition to give at this point in these notes. We shall make connections with ordinary mathematical reality shortly. Just as in Corollary 10.4, we now know that no knot with $A(K) = 1$ can be ribbon.

Theorem 10.6. Two knots K and K' are pass-equivalent if and only if $A(K) = A(K')$. In particular, any knot is pass-equivalent either to the unknot or to the trefoil knot. Any knot is pass-equivalent to its mirror image.

Proof. An orientable spanning surface for a knot K can be represented as an embedding of a standard surface of genus g, seen as a disk with attached bands (see Figure 46). Up to ambient isotopy, the embedding can be so arranged that the disk is embedded in standard form while the bands may be twisted, knotted, and linked with one another. Since the edges of a given band receive opposite orientation, passing bands

(see below) result in a pass-equivalence of the surface
boundaries.

By passing bands in this manner we can disentangle the surface
until it is a boundary connected sum of embeddings of surfaces
of genus one. Further disentanglement through band passing then
reduces each genus one surface to a surface whose boundary is
either a trivial knot or a trefoil knot. The crucial move for
this process is the twist cancellation:

The surface can be represented so that all twists are
accomplished by curls of the form  (this
follows from orientability). In such a representation, the
mirror image knot is the boundary of the surface formed by
passing all bands simultaneously as in:

Thus any knot is pass-equivalent to its mirror image.

So far we have shown that any knot is pass-equivalent to a
sum of trefoils. Let K denote the trefoil knot. Then
$K \# K \approx K \# K^! \approx 0$, the first pass because K passes to its
mirror image, the second because the connected sum of a knot
and its mirror image is ribbon, hence passes to the unknot.

Thus any sum of trefoils is reduced through passing to a single trefoil or to the unknot. This completes the proof that any knot is pass-equivalent to either a trefoil or an unknot. Since these two classes are distinguished by $A(K)$, this completes the proof of the theorem.

Remark. Part of the disentanglement process of this proof is illustrated in Figure 47.

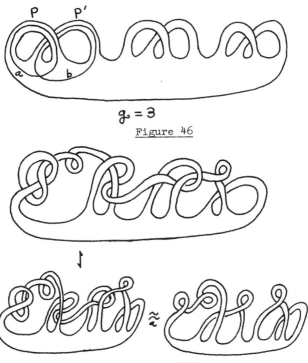

$g_o = 3$

Figure 46

Figure 47

Seifert Pairing and Arf Invariant

The Arf invariant of a knot is usually (see [34]) defined as the Arf invariant of a mod-2 quadratic form that is associated with the knot. We now give a brief outline of this version.

A mod-2 quadratic form is a function $q : V \longrightarrow Z/2Z$ where V is a finite dimensional vector space over the field of two elements $(Z/2Z)$, and q satisfies the formula $q(x+y) = q(x) + q(y) + x \cdot y$ where $x \cdot y$ is a symmetric bilinear form defined on V. The quadratic form is said to be non-degenerate if this bilinear form is non-degenerate. This, in turn, means that a matrix for the bilinear form with respect to a basis for V is non-singular. When V has even dimension the non-degenerate bilinear form is particularly simple in a symplectic basis $B = \{e_1, f_1, \ldots, e_g, f_g\}$ where

$e_i \cdot e_j = f_i \cdot f_j = 0$ for all i, j
$e_i \cdot e_j = \delta_{ij}$.

Here δ_{ij} is the Kronecker delta. Any non-degenerate form has a symplectic basis.

When q is a non-degenerate quadratic form on a vector space of even dimension 2g, then it is not possible for an equal number of elements of V to go to 0 and to 1 under q. The Arf invariant of q, $ARF(q)$, is defined to be 0 or 1 according as the majority of elements of V go to 0 or to 1 under q.

Let K be a knot in the three-dimensional sphere S^3, and let $F \subset S^3$ be an oriented spanning surface for K. Let $\theta : H_1(F) \times H_1(F) \longrightarrow Z$ be the Seifert pairing as defined in section 8 of these notes. Let $V = H_1(K; Z/2Z)$, and define $q : V \longrightarrow Z/2Z$ by the formula $q(x) = \theta(x,x)$ (mod-2). Then one can verify that q is a mod-2 quadratic form that is non-degenerate. It is associated with the mod-2 intersection form for cycles on the spanning surface. One well-known definition for the Arf invariant of the knot K is: $ARF(K) = ARF(q)$ with q defined as above.

Theorem 10.7 (<u>Folklore</u>). Let K and $\bar{K}$ be two knots that differ by switching one crossing, and let L be the link of two components obtained by splicing this crossing. Let $Lk(L)$ denote the linking number of this two-component link. Then the Arf invariants of K and $\bar{K}$ are related by the formula

$$ARF(K) - ARF(\bar{K}) \equiv Lk(L) \quad (\text{modulo } 2).$$

<u>Proof</u>. By using Seifert's algorithm for constructing a spanning surface (see section 8) plus the remarks in the proof of Theorem 10.6, we may assume the crossing being switched corresponds to a $180°$ twist in one of the bands on the spanning surface. The link L is obtained by cutting this band and taking the boundary of the resulting surface. See Figure 48. Let the band under discussion be labelled P. We may assume that it is associated with a band P' one of whose feet stand between the feet of P as in the normal form for an oriented surface (Figure 46). Let a and b be cycles

on the surface, running through the bands P and P' respectively (See Figures 46 and 48.). Then one can verify that $\theta(b,b) = Lk(L)$.

We use this fact to prove the Theorem: The Arf invariant of a quadratic form q can be computed from a symplectic basis by the formula

$$ARF(q) = q(e_1)q(f_1) + \ldots + q(e_g)q(f_g)$$

(See [34]). The curves a and b can be placed in a symplectic basis for $H_1(F;Z/2Z)$ so that $e_1 = a$ and $e_2 = b$. Here F denotes the spanning surface for the knot K. It is then easy to see that the difference of the Arf invariants for K and $\bar{K}$ is equal to $\theta(b,b)$ (modulo 2) where θ is the Seifert pairing for F. Hence, by the above remarks,

$$ARF(K) - ARF(\bar{K}) = Lk(L) \quad (modulo\ 2).$$

This completes the proof.

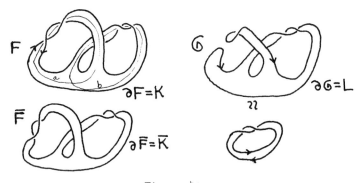

Figure 48

Corollary 10.8. Let K be a knot, and let $A(K)$ denote the mod-2 reduction of $a_2(K)$ as in Definition 10.5. Then $A(K) = ARF(K)$.

Proof. Let K and $\bar{K}$ be as in Theorem 10.7. From Lemma 5.6 and the Identities for the Conway polynomial we have the formula $a_2(K) - a_2(\bar{K}) = Lk(L)$ (assuming that K has the positive crossing). Hence $A(K) - A(\bar{K}) = Lk(L)$ (modulo 2). Since this identity suffices to calculate $A(K)$ from its value on the unknot, and since $ARF(K)$ satisfies an identical identity by Theorem 10.7, it suffices to show that $ARF(K)$ and $A(K)$ agree on the unknot. This is true.   Q.E.D.

This completes our brief description of the standard approach to the Arf invariant of a knot. We end with a short proof of Levine's Theorem (see [25]).

Theorem 10.9. Let K be a knot in the three-dimensional sphere. Let $\Delta_K(t)$ denote the Alexander polynomial of K. Then
$ARF(K) = 1 \iff \Delta_K(-1) = \pm 3$ (modulo 8).
$ARF(K) = 0 \iff \Delta_K(-1) = \pm 1$ (modulo 8).

Proof. Let $\doteq$ denote equality up to sign and powers of t. Then we have the identity $\Delta_K(t) \doteq \nabla_K(\sqrt{t} - 1/\sqrt{t})$, (Theorem 6.9). Hence $\Delta_K(-1) \doteq \nabla_K(2\sqrt{-1})$. Since K is a knot, the Conway polynomial has only even powers of z in $\nabla_K(z)$. Therefore

$$\nabla_K(2\sqrt{-1}) = 1 + a_2(K)(2\sqrt{-1})^2 + a_4(K)(2\sqrt{-1})^4 + \ldots .$$

Since $A(K) = ARF(K) = a_2(K)$ (modulo 2), we have

$$\Delta_K(-1) \doteq 1 + 4ARF(K) \quad (\text{modulo } 8).$$

This completes the proof.

APPENDIX. The Classical Alexander Polynomial

In this appendix we shall sketch one approach to the Alexander polynomial. This material is standard, and is based upon Alexander's original paper [1].

Let G be a finitely presented, finitely related group that is equipped with a surjective homomorphism $f : G \longrightarrow Z$ where Z denotes the group of additive integers. Assume that $\text{Kernel}(f) = G'$, the commutator subgroup of G. Let $H = G'/G''$ be the abelianization of this commutator subgroup. If s is an element of G such that $f(s) = 1$ and x denotes the (multiplicative) generator of the group ring $\Gamma = Z[Z] = Z[x, x^{-1}]$, then H becomes a Γ-module via the action on G': $x(g) = sgs^{-1}$.

We say that the pair (G,f) is an __indexed group__. Two indexed groups (G_1, f_1), (G_2, f_2) are isomorphic if there is a group isomorphism $h : G_1 \longrightarrow G_2$ such that $f_2 \circ h = f_1$. If $H(G,f)$ denotes G'/G'' with module structure as above, then $H(G,f)$ is an isomorphism invariant of the indexed group (G,f).

As we shall see, $H(G,f)$ is finitely presented and related over Γ. Suppose that there exists such a presentation with an equal number n of generators and relations. (This will be the case for the knot group.) Then there is an exact sequence

$$\Gamma^n \xrightarrow{A} \Gamma^n \longrightarrow H(G,f) \longrightarrow 0$$

where A is an $n \times n$ matrix with entries in Γ.

Let $D(A)$ denote the determinant of A. The Γ-module structure on $H(G,f)$ is induced by scalar (Γ) multiplication on Γ^n. Let $\hat{A}$ denote the adjoint matrix to A so that $A\hat{A} = D(A)I$ where I is the $n \times n$ identity matrix. This shows that for all a, $D(A)a = A(\hat{A}a)$, and hence $D(A)a$ is in the image of A. Therefore $D(A)[a] = 0$ for all $[a] \in H(G,f)$. Thus $D(A)$ is an annihilating element for $H(G,f)$ as a Γ-module. We shall see that the Alexander polynomial takes the form of $D(A) = \Delta_K(x)$ for an approrpriate matrix A.

Before doing more algebra, let's turn to the geometry. Let $G = \pi_1(S^3-K)$ be the fundamental group of the knot complement. This group is finitely presented and related, with a particularly useful presentation known as the Dehn presentation. In the Dehn presentation each region of the knot diagram corresponds to an element of $\pi_1(S^3-K)$ via the following conventions: Replace S^3-K by R^3-K and assume that the knot lies in the $(x,y,0)$ plane, except for over and under-crossings. These crossings deviate in the z-direction (third variable) by $0 < |z| \ll 1$. Let the basepoint $p = (0,0,1)$ be a point above the knot diagram. Associate to each region R a loop that starts at p, descends to pierce R once, and then returns by piercing the unbounded region once. See Figure 49.

In this form each region of the knot diagram corresponds to a generator of the fundamental group, except for the unbounded region, which corresponds to the identity element.

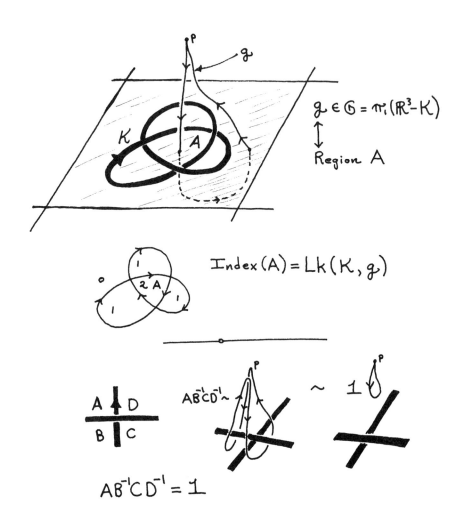

Crossing relation in Dehn presentation

Figure 49

Each crossing in the knot diagram corresponds to a relation in the fundamental group (as illustrated in Figure 49). This gives a complete set of relations for the group. The mapping $f : \pi_1(S^3-K) \longrightarrow Z$ exists since a fundamental group of a knot complement abelianizes to Z. The map can be specifically described by linking numbers: $f([g]) = Lk(K,g)$ where Lk denotes linking numbers of curves in R^3. With this interpretation we see that, when we let elements of the group correspond to regions R in the knot diagram as described above, then $f(R) = \text{Index}(R)$ where Index denotes the Alexander index of the region R (See Lemma 3.4.). (The unbounded region is assigned index zero. See Figure 49.)

Remark. The Γ-module $H(G,f)$ has the following interpretation. Let $q : X \longrightarrow S^3-K$ be the covering space corresponding to the representation $f : G \longrightarrow Z$. The space X is the infinite cyclic cover of the knot complement (see [26]). The first homology group, $H_1(X;Z)$, is a Γ-module via the action of the group of covering translations of X. With this structure, $H_1(X;Z)$ and $H(G,f)$ are isomorphic Γ-modules. This interpretation is very important, but will not be pursued here.

Returning to algebra, we wish to describe a presentation for G', and thence compute G'/G''. Suppose that G has a presentation of the form: $G = (s,g_1,g_2,\ldots,g_n/R_1,\ldots,R_m)$ with

1. $n = m$ (True for the Dehn presentation since there are two more regions than crossings, and one region corresponds to the identity element.) or $m \geq n$.

2. $f(s) = 1$, $f(g_1) = f(g_2) = \ldots = f(g_m) = 0$.

The second condition is accomplished from an arbitrary presentation by choosing an s ($f(s) = 1$), and re-defining the other generators via multiplication by approrpriate powers of s, to insure that they all hit zero under f.

Recall that $\Gamma = Z[x, x^{-1}]$ acts on G via $xg = sgs^{-1}$. It is easy to see that G' is generated by the set $\{x^k g_i / k \in Z, i = 1, \ldots, n\}$. In particular, each relation R_k can be rewritten in terms of these generators. Let $\rho(R_k)$ denote this rewriting of R_k. Then $\{x^k \rho(R_k) / k \in Z, i = 1, \ldots, m\}$ is a set of relations for G' (proof via covering spaces or combinatorial group theory). Thus $G' = (\{x^k g_i\} / \{x^k \rho(R_j)\})$. By abelianizing these generators and relations, and writing them additively, we obtain the structure of $H(G, f)$.

Consider the Dehn presentation. Let s correspond to a region of index 1 (or -1 if necessary). Suppose A, B, C, D are the regions around a crossing with indices $f(A) = p$, $f(B) = f(D) = p+1$, $f(C) = p+2$ as in Figure 50. Then we have new generators a, b, c, d with $A = s^p a$, $B = s^{p+1} a$, $C = s^{p+2} c$, $D = s^{p+1} d$. The relation $R = AB^{-1}CD^{-1}$ becomes

$$R = AB^{-1}CD^{-1}$$
$$= (s^p a)(s^{p+1} b)^{-1}(s^{p+2} c)(s^{p+1} d)^{-1}$$
$$= s^p ab^{-1} scd^{-1} s^{-p-1}.$$

The next calculation rewrites R in terms of the generators $x^k a$, $x^k b$, $x^h c$, $x^k d$ of G'.

$$R = s^p a s^{-p} s^p b^{-1} s^{-p} s^{p+1} c s^{-p-1} s^{p+1} d^{-1} s^{-p-1}$$
$$= (x^p a)(x^p b)^{-1}(x^{p+1} c)(x^{p+1} d)^{-1} \equiv \rho(R).$$

Upon abelianizing G' to form $H(G,f)$, this relation becomes the additive relation $x^p(a - b + xc - xd)$. Thus, as a module over the group ring Γ, $H(G,f)$ is generated by symbols $a, b, c, d, \ldots$ corresponding to the regions of the knot diagram, with generating relations, one per crossing, of the form $a - b + xc - xd$. These relations can be remembered by placing ± 1, $\pm x$ around the crossing in the regions that they correspond with (as shown in Figure 50).

Figure 50

Earlier in the notes we have referred to this labelling as the Alexander code. The symbol corresponding to the unbounded region is set equal to zero, and an adjacent region corresponds to s ($f(s) = 1$) and is also eliminated. The resulting square relation matrix is exactly what we have described in section 3 as the Alexander Matrix for this code. Its determinant is the Alexander polynomial, $\Delta_K(x)$.

That the Alexander polynomial is well-defined up to sign and powers of x and that it is a topological invariant of the knot K can be verified by examining the combinatorial group theory that we have sketched.

Alexander apparently felt that the algorithm should stand on its own right, and he wrote his first paper on the polynomial from a combinatorial standpoint with slight mention of the background group theory and topology.

Remark on Determinants

In these notes we have used a formulation of certain determinants (of Alexander matrices) as state summations over the states of a universe (knot or link graph). The signs in the determinant expansion come from the geometry of the universe. I wish to point out here that the formula for any determinant follows a similar pattern.

The sign of a permutation is determined geometrically by the following prescription. List the numbers $1,\ldots,n$ in order, and the permutation of them on a line below. Connect corresponding numbers by arcs so that all arcs intersect transversely at double points. Then the sign of the permutation is $(-1)^c$ where c is the number of intersections of the arcs. For example let $P = \begin{pmatrix} 1 & 2 & 3 & 4 & 5 & 6 \\ 6 & 3 & 4 & 2 & 1 & 5 \end{pmatrix}$:

The combinatorics underlying the determinant expansion of an $n \times n$ matrix consists of the $n!$ <u>grid states</u> of an $n \times n$ grid. A grid state is a pattern of n rooks on the $n \times n$ chessboard so that no two rooks attack each other. (Rooks move only on horizontal and vertical files. Thus no file contains more than one rook.) For example, the grid states for a 2 x 2 board are  .

Given a grid state S and matrix A, define $\langle A|S\rangle$ to be the product of the entries of A that are in boxes corresponding to rooks of S (when the matrix and grid state are superimposed).

The sign of a grid state S, denoted $\sigma(S)$, is the sign of the permutation of the rows that produces this state from the diagonal state (all rooks on the main diagonal). To obtain this sign directly, draw arcs outside the grid from top row positions to left column positions so that each arc marks the row and column position of a corresponding rook. Then $\sigma(S) = (-1)^c$ where c is the number of crossings of the arcs. Then $\text{Det}(A) = \sum_{S} \sigma(S)\langle A|S\rangle$ where $\mathcal{S}$ is the collection of all grid states for the n x n grid.

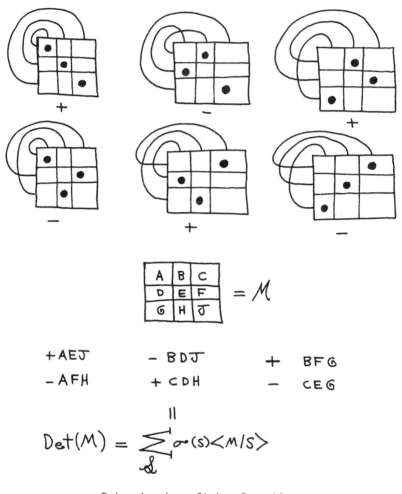

Determinant as State - Summation

Figure 51

REFERENCES

1. J.W. Alexander. Topological invariants of knots and links. Trans. Am. Math. Soc., 30 (1923), 275-306.

2. R. Ball and M.L. Mehta. Sequence of invariants for knots and links. (To appear in Journal de Physique.)

3. J. Birman and R.F. Williams. Knotted periodic orbits in dynamical systems I: Lorenz's equations. Topology 22 (1983), 47-82.

4. A. Casson and C. McA. Gordon. On slice knots in dimension three. In Geometric Topology, R.J. Milgram (editor). Proceedings of Symposia Pure Mathematics XXXII, Amer. Math. Soc., Providence 1978, 39-53.

5. J.H. Conway. An enumeration of knots and links and some of their algebraic properties. Computational Problems in Abstract Algebra. Pergamon Press, New York (1970), 329-358.

6. J.H. Conway. Lecture, University of Illinois at Chicago. Spring 1978.

7. Daryl Cooper. Ph.D. Thesis. Mathematics Institute, University of Warwick (1981).

8. R.H. Crowell. Genus of alternating link types. Ann. of Math. 69 (1959), 258-275.

9. R.H. Crowell. Nonalternating links. Ill. J. Math. 3 (1959), 101-120.

10. R.H. Crowell and R.H. Fox. Introduction to Knot Theory. Blaisdell Publishing Company (1963).

11. L. Euler. Solutio problematis ad geometriam situs pertinentis. Comment. Academiae Sci. I. Petropolitanae 8 (1736), 128-140. Oper Omnia Series 1-7 (1766), 1-10.

12. R.H. Fox. A quick trip through knot theory. Topology of Three Manifolds. Prentice Hall (1962), 120-167.

13. R.H. Fox and J.W. Milnor. Singularities of 2-spheres in 4-space and cobordism of knots. Osaka J. Math. 3 (1966), 257-267.

14. C.F. Gauss. Zur geometrie der lage fur zwei raum dimensionen, Werke, Band 8, 272-286, Konigl. Gesellshaft der Wissenshaften, Goetigen, 1900.

15. C. Giller. A family of links and the Conway calculus. (To appear.)

16. I. Handler. (Private communication.)

17. R. Hartley. The Conway potential function for links. (To appear.)

18. L.H. Kauffman. Link manifolds. Mich. Math. J. <u>31</u> (1974), 33-44.

19. L.H. Kauffman. Planar surface immersions. Ill. J. Math. <u>23</u> (1979), 648-665.

20. L.H. Kauffman. Weaving patterns and polynomials. Topology Symposium Proceedings, Siegen 1979. Springer Verlag Lecture Notes in Mathematics <u>788</u>, 88-97.

21. L.H. Kauffman. The Conway polynomial. Topology <u>20</u> (1980), 101-108.

22. L.H. Kauffman and T.F. Banchoff. Immersions and mod-2 quadratic forms. Amer. Math. Monthly <u>84</u> (1977), 168-185.

23. L.H. Kauffman and L. Taylor. Signature of links. Trans. Amer. Math. Soc. <u>216</u> (1976), 351-365.

24. J. Levine. Knot cobordism groups in codimension two. Comm. Math. Helv. <u>44</u> (1969), 229-244.

25. J. Levine. Polynomial invariants of knots of codimension two. Ann. of Math. <u>84</u> (1966), 534-554.

26. J.W. Milnor. Infinite cyclic coverings. <u>Topology of Manifolds</u>. (Mich. State Univ. 1967). Prindle Weber and Schmidt, Boston (1968).

27. J.W. Milnor. <u>Singular Points of Complex Hypersurfaces</u>. Princeton University Press (1968).

28. K. Murasugi. On the genus of the alternating knot I. J. Math. Soc. Japan <u>10</u> (1958), 94-105.

29. K. Murasugi. On the genus of the alternating knot II. J. Math. Soc. Japan <u>10</u> (1958), 235-248.

30. K. Murasugi. On the Alexander polynomial of the alternating knot. Osaka J. Math. <u>10</u> (1958), 181-189.

31. K. Murasugi. A certain numerical invariant of link type. Trans. Amer. Math. Soc. <u>117</u> (1965), 387-422.

32. K. Murasugi and E.J. Mayland, Jr. On a structural property of the groups of alternating links. Can. J. Math. <u>28</u> (1976), 568-588.

33. K. Reidemeister. **Knotentheorie**. Chelsea Publishing Company, New York (1948). Copyright 1932. Julius Springer, Berlin.

34. R. Robertello. An invariant of knot cobordism. Comm. Pure Appl. Math. 18 (1965), 543-555.

35. D. Rolfsen. **Knots and Links**. Publish or Perish Press (1976).

36. H. Seifert. Uber das geschlecht von Knoten. Math. Ann. 110 (1934), 571-592.

37. J. Stillwell. **Classical Topology and Combinatorial Group Theory**. Springer Verlag (1980).

38. H. Whitney. On regular closed curves in the plane. Comp. Math. 4 (1937), 276-284.

Department of Mathematics, Statistics and
    Computer Science
The University of Illinois at Chicago
Chicago, Illinois   60680

Library of Congress Cataloging in Publication Data

Kauffman, Louis H., 1945–
  Formal knot theory.

  (Mathematical notes ; 30)
  Bibliography: p.
  1. Knot theory. I. Title. II. Series: Mathematical
notes (Princeton University Press) ; 30.
  QA612.2.K37 1983   514'.224   83-42594
  ISBN 0-691-08336-3 (pbk.)

  Professor Kauffman is Associate Professor of Mathematics
at the University of Illinois at Chicago.